SpringerBriefs in Education

For further volumes:
http://www.springer.com/series/8914

Virginia McShane Warfield

Invitation to Didactique

 Springer

Virginia McShane Warfield
Department of Mathematics
University of Washington
Seattle, WA, USA

ISSN 2211-1921   ISSN 2211-193X (electronic)
ISBN 978-1-4614-8198-0   ISBN 978-1-4614-8199-7 (eBook)
DOI 10.1007/978-1-4614-8199-7
Springer Tokyo Heidelberg New York Dordrecht London

Library of Congress Control Number: 2013945307

Printed on acid-free paper

Springer is part of Springer Science+Business Media (www.springer.com)

Foreword

Half a century ago, the French word *didactique,* like its counterpart *didactic* in English, was an adjective with two meanings. The first was the neutral "intended to instruct," and the second was the pejorative "overly inclined to lecture others." Today, through the efforts of French researchers in mathematics education, *didactique (des mathématiques)* has been given a third meaning: as a noun denoting a field that studies questions raised by teaching and learning (of mathematical knowledge) in the milieu of school. *Didactique* has expanded during the past several decades across national borders and, through many translations, beyond the French-speaking community of mathematics educators. Nonetheless, despite a rich and growing literature base, it has as yet had only modest influence in the Anglophone world.

Ginger Warfield has played a pivotal role in translating into English much of the pioneering work in the field of *didactique* over the past decades. In the volume at hand, she introduces the reader with an interest in mathematics education but unfamiliar with the abstruse literature of *didactique* to basic ideas of the field coming from its genesis in the theory of didactical situations. Resonating in this book is the lucid, knowledgeable voice of one who appreciates the complexity of *didactique* while being able to express in compelling language its power and promise. Welcome to a realm of mathematics education that you may have only glimpsed fragments of until now but that represents an extensive share of the best thinking we have concerning how mathematics is and can be taught.

Athens, GA, USA   Jeremy Kilpatrick

Preface to the 2013 Edition

The 6 years since this book was first published have been fruitful ones both for Guy Brousseau, whose work is at the core of the book, and for our partnership. Having retired a few years back, he has set himself the task of making his work as accessible as possible to people who are interested. One of the results of this has been sending a truckload of artifacts (student papers, meeting notes, etc.) from the Michelet School, where much of his research occurred, to the library of the Université Jaume 1 in Castelo, Spain, where they are being made available to researchers. Another has been helping to launch the ViSA site (Vidéos de situations d'enseignement et d'apprentissage), a collection of videos on teaching and learning (http://visa.inrp.fr/visa), and to set up there many of the videotapes of the work from the Michelet School that has served, and continues to serve, as a basis for research.

A third element, which is a work in progress, is his website, http://guy-brousseau.com, onto which he is uploading a lifetime's worth of articles and Power Point presentations and other relevant items. Since his site is, of course, in French, and since machine translation produces absolutely hair-raising results, I am creating an English language mirror site, http://faculty.washington.edu/warfield/guy-brousseau.com. I will not even attempt to provide translations of all of the hundreds of entries on the original site, but I have translated some key pages and intend to continue translating the ones that strike me as most useful. All of the links remain, but you never know when you click whether you will wind up in French or English.

And on the hard copy front, the series of four Brousseau, Brousseau, and Warfield articles on "Rationals and Decimals as Required in the School Curriculum" have now all been published in the *Journal of Mathematical Behavior*. After we finished them, the three of us realized that together they constituted the core of a book that would have even more to say. Writing the book showed us how right we were, and opened up whole new topics that had never been written up and could add a great deal to readers' understanding of the origins and objectives of *didactique* and of the Theory of Situations—in fact, of all of Brousseau's work. That book is now in press with Springer, with the title *Teaching Fractions Through Situations: A Fundamental Experiment*.

One consequence of the writing of the book has been that my own understanding of and fascination with the field have deepened. It was therefore with some trepidation that I picked up this book to check it over for this edition. To my relief, nothing revealed itself as false. There are nuances and depths missing, but then again the book has always been intended as a primer, not an in-depth study.

Only one glaring gap leapt to my eye, and I shall take this opportunity to fill it: in 2003, Guy Brousseau was awarded the first Felix Klein Medal of the International Commission on Mathematical Instruction, recognizing his outstanding lifetime achievements in mathematics education research. For me, this was doubly positive. On the professional front, it was a confirmation of the belief I had long held of the importance of Brousseau's work. And on a personal front, it was a joy that international recognition was being given to someone for whom my great respect is augmented by a great affection.

Preface to the 2007 Edition

Five or six years ago, at a meeting of the American Educational Research Association, Carolyn Maher and I had what turned out to be a high-impact conversation. We were discussing the work of Guy Brousseau—how much we both liked it, and how frustrated we were that the book that had been published in 1996 was proving difficult of access to many genuinely interested English speakers. "Someone," she said, fixing me with a penetrating eye, "needs to write a short, easy-to-read monograph that provides the information needed for reading the book itself. I would see to it that it got published." It was immediately clear to me that she was right. It was also clear to me, given that I have had the great privilege of working closely with Brousseau for several years, who the "someone" needed to be.

This book has been a work in progress ever since. Various phases of it have received many helpful comments and suggestions from Carolyn and her coeditor Bob Speiser. In addition, Susan Pirie has at least twice caused me to dismantle whole chapters and reassemble them in drastically improved form. In the end, it became clear that publication through conventional channels was no longer the appropriate format for achieving our original goal of swift accessibility, and so I opted for publishing it with X-libris instead. The change in format did not, however, change at all the fact that I am very grateful to all three of them.

Contents

Introduction

If a picture is worth a thousand words, sometimes an example is worth a thousand definitions. Before beginning to delve into the history, development, and content of the Theory of Didactical Situations, let us examine a basic example of a Situation.

An Early Counting Situation

Setup: On one side of the classroom is a table on which there is a set of paint pots (three or four of them). On the other side is a collection of paint brushes.

Objective: For children for whom counting is still fairly novel to create ways in which to communicate in writing the results of their counting.

Activity: One child, equipped with paper and a pencil, is to go to the paint pots and write on the paper a message of some form (any form) that will enable another child to know how many pots there are. The paper is then to be given to a second child, who is to go to the brush collection and use the message to decide how many brushes to take. The second child is then to take the brushes to the paint pots and put one brush in each pot. If the message was correct, and correctly interpreted, then there will be one brush in each pot and no brushes left over.

The first thing that leaps to the eye is what fun very young children could have with this activity. The next is how little interference it requires from the teacher. As soon as they know what to do, the children also know how to tell whether they have succeeded. As successive pairs try it out, they can experiment with different "codes." Depending on the level of sophistication of the particular child, a code might be a drawing of the paint pots, or a tally, or the numeral itself. By examining other children's messages, children expand their own repertoire of codes and can compare their merits.

A little further thought adds the realization that the teacher, relieved of the responsibility of directing activities, can be a very close observer. Watching the children's processes of solution and interactions provides the teacher with a far better assessment of what each child understands or is struggling to learn about numeration and counting than a personal interrogation could do.

Reflecting a little further, one notes that this is also an adjustable activity. For a group or an individual not yet ready to try the communication game, the solo version may be more appropriate and highly effective: "Go look at the pots, then in a single trip collect enough brushes to put exactly one in each pot." For a more advanced group, augmenting the number of pots or elaborating the directions ("put two in each pot") could be interesting.

More cogitation reveals how effective this one activity is. It brings out whether a child actually knows what counting is. Many an observant parent will attest that children tend first to learn the "song" ("one, two, three, four, six, nine"), then gradually, while perfecting the song, pick up the idea of establishing a one-to-one correspondence between a set of objects and the numbers in order. With that accomplished, the next task is to learn to make use of that correspondence. A classic example of a child who had reached the former stage but not the latter is the story of a little boy in the pots of paint Situation whose classmates tried to help him past his difficulties with it by whispering "Count! Count!" Obediently and correctly he counted every brush, then grabbed a random handful and headed back toward the pots.

What no amount of thinking in the abstract about this Situation can reveal is the extent to which it has been used for research purposes in *Didactique*. It has given information on the difference between the grasp on numbers of a child functioning in isolation and one functioning as part of a group. It has provided some enlightenment on how to work around the fact that for almost all children there is a radical difference in relationship between numbers less than five (which they grasp simply as a characteristic of a collection in the same way that they grasp color or texture) and numbers above six. As a representative of a type of Situation, it has provided a means for studying the functioning of Situations themselves — what the impact is of a change in each of its variables (including, for instance, exchanging the paint pots and brushes for little animals to be rounded up and moved around on a computer screen). With a slight increase in the level of sophistication, it becomes an instrument for the study of how children cope with numbers slightly beyond their level of counting competence.

The example above provides a hint of the depth of *Didactique*. It also provides an opportunity to describe some of the hazards encountered in an attempt to communicate it to the English-speaking world. For one thing, *Didactique* has been a research program unifying a considerable community of mathematics educators in France for upwards of thirty years, but very little information about *Didactique* has penetrated the French/English language barrier. As a result, it has developed a culture and vocabulary which, while they clarify communication within the community, tend rather to obscure it when viewed from the outside. This style and use of terminology can in itself be a definite obstacle to the comfort of the reader.

The other hazard to communication has a very different flavor. Many of the Situations, like the one above, are based on activities that are very attractive to teachers. There is a major temptation to go through the field and mine it for these nuggets. But in fact, the current international scene in mathematics education contains a plethora of interesting and locally valuable mathematical activities.[1] If *Didactique*'s Situations are stripped of their context and if their relationship to each other is hidden from sight, then the obvious—and totally reasonable—response is "That's very nice, but what's all the fuss about?"

Having now provided an appetizer as well as a couple of warnings, I propose in the first chapter to offer the reader a two-stage introduction to the area of *Didactique* known as the Theory of Situations. This in turn will situate the reader who wishes to study other aspects of *Didactique*, since, in the words of Marie-Jeanne Perrin-Glorian, "...most of the research [in *Didactique* since the sixties] has used the theory of didactical Situations, or positioned itself relative to it, or asked it questions or even contributed to its evolution" [Perrin-Glorian, 1994]. The first stage provides a historical orientation. Since the theory originated with Guy Brousseau, an understanding of his intellectual and academic sources can be extremely enlightening, enabling people unfamiliar with the French system to see the coherence of the theory and to place it in a reasonable context. The orientation thus consists of a mixture of biographical material with which he has kindly provided me, and information from the article by Marie-Jeanne Perrin-Glorian quoted above.

The second stage takes a selected subset of the major concepts of *Didactique* and provides definitions, illustrations, and discussions of them. The intention is to provide the reader with a firm enough grasp on these concepts so that they will be welcome landmarks in future reading. On the other hand, since they will not be discussed in their full splendor, references will also be provided (most notably in Brousseau's Theory of Didactical Situations).

The second chapter, by way of further illustration, provides a more detailed look at one particular sequence of Situations. The third and fourth chapters summarize articles dealing with two of the major concepts of *Didactique*: Fundamental Situation and didactical contract. The fifth chapter is written in the hope that this book has succeeded in whetting the reader's appetite for *Didactique*. It is a glossary designed to aid in the reading of other documents in the field.

One further note: I would like to emphasize an aspect that should be clear by now. This is in no way a survey of *Didactique*. It is an introduction, inviting the reader to learn about *Didactique* by following one particular route into it, namely, by following at least to some extent in the steps of Guy Brousseau. This is not unreasonable, given that he originated the field and is still working in it. On the other hand, many other people have joined him and done very interesting work. It is only in the interest of avoiding excess volume and density that their work is omitted here.

---

[1] For instance, those found in the NCTM's *Teaching Children Mathematics, Mathematics Teaching in the Middle School*, and *Mathematics Teacher,* ARDM's *Grand N* and *Petit x,* and the Freudenthal Institute's *Nieuwe Wiskrant.*

Chapter 1
Basic Elements of *Didactique*

Background and Development of the Field

Guy Brousseau was born in Morocco in 1933. Since his father was in the military, he wound up attending school in a number of different places. He was nonetheless quite successful, and earned himself admission to the *École Normale d'Instituteurs* (Normal School for Elementary Teachers) in Agen. He earned a baccalaureate degree there, and then a more advanced baccalaureate (with distinction in elementary mathematics) at the *École Normale* in Montpellier. This in turn led to his being given a scholarship to study mathematics in Toulouse in preparation for being admitted to an *École Normale Supérieure*. By the end of the year, however, he realized that this was not the direction he wanted to go. Much as he enjoyed the advanced mathematics and physics he was learning, his real fascination was with how children learn mathematics—not how children learn in general but very specifically how they learn mathematics. When one of his professors suggested that he belonged in the field of psychology he rejected the notion out of hand. In fact, even more specifically, he wanted to focus on how students learn mathematics in a classroom. He had, by this time, been exposed to the work of Piaget, and greatly admired many of his ideas and explorations, but felt that Piaget's focus on individual children isolated from any group excluded some key learning dynamics. In pursuit of this fascination, he returned to Agen for a year to complete the studies required for a teaching certificate. At the end of the year he was given a position in a village school, teaching all of the village's elementary age children. Shortly thereafter he married Nadine Labesque, whom he had met on his first day in Agen, and as a couple they were given a school in a slightly larger village. Nadine taught the younger half of the children, Guy the older. These years launched his experimental efforts, both in the classroom (his and Nadine's) and out of it (for instance teaching a collection of farmers techniques for optimization and for computing areas.) He also continued learning about other people's ideas by spending every spare moment in the nearby Tonneins or not-too-distant Toulouse. He gives a vivid description of himself standing in a bookstore voraciously consuming books (he bought enough to

V.M. Warfield, *Invitation to Didactique*, SpringerBriefs in Education 30,
DOI 10.1007/978-1-4614-8199-7_1, © The Author(s) 2014

keep from being thrown out!) To avoid wasting any of his precious time, he developed a practice of invariably starting a book in the middle, so as to determine whether it had anything of interest and use to him. If he decided that it did, he generally figured out on his own what must have come before, only rarely actually going back and reading the early parts. In this way he acquired extensive knowledge of a wide range of material with remarkably little exposure to how the material is introduced to the more conventional reader who starts on page one.

This phase lasted from 1954 to 1956, at which point he was called up for military service. A high score on an examination early in his service resulted in his being able to choose further training in Paris. In his spare time he enrolled for a course at the Sorbonne in which he expected to review his measure theory. Instead it turned out to be one of the early expositions of "Modern Math". This was part of a project launched by a number of young French mathematicians who published under the collective pseudonym of Bourbaki. Their goal was to break what they saw as a detrimentally tradition-bound pattern of teaching mathematics at the advanced levels (the "Grands Écoles") by introducing a new structure with set theory at its foundation. He found the ideas exciting, though he swiftly realized that the structure for that level didn't coincide with the structure he was interested in producing at the elementary level.

For the rest of his time in the army he was in Algiers. His duties as Officer in charge of Transmission gave him time to pursue his ideas and produce multitudinous experimental problems and worksheets. He returned to civilian life and his family and classroom in January of 1959. For the next several years he taught, formulated ideas that he and Nadine then tried out in their respective classrooms, and wrote his ideas and results in gray notebooks (many, many gray notebooks!). He also continued to read extensively. One of his readings in 1961 had a dramatic impact: a book by Mme. Lucienne Félix agreed so completely with his ideas on teaching and learning that he decided to write to her. Lucienne Felix was an able mathematician whose gender and religion (she was Jewish) severely limited her formal career. Undeterred, she taught school in Paris and became passionately involved with issues of mathematics education. She worked closely with many noted French mathematicians who were part of France's "Modern Math" movement, such as Emile Picard, Emile Borel and Henri Lebesgue. In particular, she became an assistant to Lebesgue, and wrote a number of articles about his work. When she read Brousseau's ideas, she in turn was very much impressed with what he sent her and immediately asked for more. She took him under her wing, and using her knowledge of and contacts with the community of mathematicians interested in education, saw to it that he was invited soon after to a conference of the International Commission for the study and improvement of the teaching of mathematics[1] where he met many members of that community. This mentor and supporter relationship developed into a profound friendship with the whole Brousseau family that lasted until her death in 1994. Recently Brousseau completed the circle by publishing her autobiography.[2]

---

[1] *Commission Internationale pour l'étude et l'amélioration de l'enseignement des mathématiques* (CEIAEM).

[2] Lucienne Félix, *Réflexions d'une Agrégée de Mathématiques au XXᵉ Siècle*, L'Harmattan, 2005.

In the fall of 1962, Brousseau won a fellowship to go to the University of Bordeaux. For the rest of the 1960s he was financially supported by a series of bureaucratic arrangements which boggle the non-French mind, but which served admirably to give him the scope he needed to develop his ideas and build the kind of research structure these ideas required. An initial step was the setting up of a Center for Research on the Teaching of Mathematics[3] within the existing Regional Center of Pedagogic Documentation.[4] This enabled him to assemble a collection of people from the University and the Normal School to work with him on publishing many worksheets and sequences of problems. One of the purposes was to help teachers by suggesting innovations and ways of teaching. Another was to study and make explicit the conditions required for doing scientific research on the subject of the teaching of mathematics. He delved deeply into this latter, studying technological, sociological and pedagogical conditions, and conferring with a wide variety of people.

The ambition he formed in the period was straightforward, if a bit breath-taking: he wanted to re-design the entire French elementary mathematics education system. The first step was to determine in a scientific manner what would be best. This involved examining the whole body of mathematics whose teaching was required by the National Programme, looking into the history of how that mathematics had been taught in the past (the past sometimes including the Greeks and Egyptians) and deciding on a which concepts led into which others. The next step was to set up a sequence of lessons shaped by that ordering and test the success rate of the children studying the sequence. The major search was for a construction of the mathematics that would also satisfy constraints imposed by pedagogy and psychology. Combined with this was a constant search for rigor in the research process.

One outcome of this project was the publication of a text book for first grade teachers, whose significance can only be understood in context. At that time, teaching in all the schools in France was governed by the National Program, published in Paris and revised periodically. All of school mathematics (and other school subjects as well) was subdivided into areas to be covered each month, and inspectors visited the schools to see that this was happening. In addition, a manual was published suggesting ways of handling the material that were consistent with the Program. Use of the manuals was not required, but it simplified life a fair amount, especially given that the manuals just happened to divide each month's worth of materials into the same number of sections as there were school weeks in that particular month. Brousseau's "Pre-school mathematics, module 1" bore no similarity to the manuals, or to the standard textbooks derived from them. The fact that the book was published by a major publisher resulted both from his powers of persuasion and his connections with Mme. Lucienne Félix and others. It served later to give him credibility in circles where he might otherwise not have had it.

The structure of the book was firmly based on the construction of a foundation for future learning of mathematical concepts, but also on connecting the mathematics with real life. The latter was very much a trend of the times. It was the period

---

[3] *Centre de Recherche sur l'Enseignement Mathématique* (CREM).
[4] *Centre Régional de Documentation Pédagogique* (CRDP).

when mathematical "manipulatives" were coming into vogue. Brousseau saw their value, but also their dangers. In particular, the prevailing assumption was that a child's skill in handling situations which an adult could see as isomorphic to a mathematical concept would transfer, so as to supply the child with the concept itself. This he maintained to be impossible unless an explicit bridge is built. In the book's introduction, he described clearly the stages in which a child's understanding (or "acquisition") of a concept proceeds, from manipulation of an object through progressively more abstract representations of the process involved. At the same time he emphasized that whereas one should not consider full understanding to have been arrived at from the moment the child succeeds with the initial manipulations, one should also not disregard the amount of understanding that has been arrived at each level just because the child can't yet verbalize the understanding.

One of Brousseau's huge assortment of activities during the 1960s was to work with a collection of teachers from several schools in the area who were interested in experimenting with the ideas represented by the book. They made worksheets modeled on these ideas, gave them to their classes, made records of the results and gathered regularly to discuss them and produce more. For Brousseau this was doubly fascinating. One element was seeing how other people responded to his ideas, what they made of them, and what the resulting impact was when they experimented with them in the classroom—an experimentation that was greatly facilitated by the fact that one of the members of the group was the inspector for the schools in question. Another element was the development of Brousseau's own ideas with regard to research on education: what was desirable, what was possible, and how to arrange it. He was very much concerned that it be conducted as experimental research with uncompromising scientific standards. One absolute requirement was objectivity, another was measurable objectives and a measurement tool whose results were independent of the person doing the measuring.

By 1968, when a national colloquium met to discuss innovations and research on teacher education which had been going on in a number of regions, there was a general agreement that there were three major objectives:

1. Fundamental research: the set of scientific investigations which make it possible to explore unknown or ill-understood domains, with no explicit practical goal;
2. Oriented research: work aimed at starting with a specific, existing situation and attaining a general objective for the teaching of mathematics; and
3. Applied research: systematic studies of knowledge, methods, techniques and instruments with precisely formulated research objectives.

Also important was the study of how to disseminate the knowledge thus developed.

Not long after the colloquium, there followed the establishment of several *IREM*[5]s—Institutes where faculty members from universities and schools could work together to do research on mathematics education. Bordeaux, thanks to the community Brousseau had built up, was admirably prepared for the situation of

---

[5] Institute pour Recherche en Éducation en Mathématiques.

such an *IREM*, and was in fact the second university to have one funded. This was a great and exciting step forward and a necessary condition for carrying out the research Brousseau had in mind. It was not, however, in his view sufficient. To fill in the gap, he threw himself into the project of creating a *COREM*[6]—an observation center for research on the teaching of mathematics. This proved to require not only a great deal of energy and patience, but also considerable tact and diplomacy, since he had to reject several generous offers from schools who wanted, for instance, to be simultaneously an observation school and a model school, or a school for training teachers. Brousseau wanted a school with a completely unselected student population for whom all the non-mathematical parts of the school day would be absolutely normal. Eventually he found just the situation he wanted at the École Michelet, an elementary school in a primarily blue collar neighborhood in the town of Talence on the outskirts of Bordeaux. He set up on its grounds an observation center, which was a detached classroom with good lighting and multiple video cameras surrounding an ordinary set of desks and blackboards. Mathematics normally happened in the regular classroom, but from time to time a class would be held in the observation center—often enough so that the children took the shift completely in their stride. He also arranged that the school would have slightly more teachers than it would normally have been allotted, so that taking part in experiments did not form an impossible burden. Nadine Brousseau, Guy's wife, was one of the people hired to fill these extra positions. She made invaluable contributions to the research both by carrying out experiments and by recording their progress, especially in the case of some of the long-term experiments.

With these tools, the research developed rapidly. In Bordeaux, the focus was increasingly on fundamental research and methodology, but the field of study enlarged. Still firmly committed to the construction of a scientific theory, it expanded from the search for a scientifically defined mathematical program to the study of the conditions of the act of teaching. *Didactique* interested itself in everything to do with the acquisition of mathematical knowledge, including the knowledge about that acquisition which other fields contribute. By 1975, Brousseau had expanded the definition of *Didactique* to (a definition presented at an inter-*IREM* colloquium):

> ...a more global *problématique* including a reflection on the purpose of teaching, on the nature of the goal-knowledge, on the methods of acquisition specific to the people who are being taught, on the objectives, on the particular theoretical and practical conditions of the pedagogical activities in the teaching of the discipline. If there is to be a *scientific Didactique*, it is necessary that it permit the deduction of the methodological measures best adapted for provoking acquisitions of knowledge from a scientific knowledge of the process of formation of the intellect (furnished in part by psychology, epistemology, biology, genetic epistemology, mathematics, linguistics, etc.) Thus it must be declared that the field of *Didactique* includes all the specific combinations of knowledge — even knowledge in other domains — which make possible the resolution of the didactician's problem: 'how to lead the student to acquire some particular notion.'

---

[6] Centre d'Observations et de Recherches sur l'Enseignement des Mathématiques [Center for Observation and Research on Mathematics Teaching].

At the same colloquium, Brousseau observed that the work of the teacher, like that of the didactician, requires that the notions to be acquired be translated in terms of behaviors expected of the students. This translation requires one to consider what it means to acquire a notion—i.e., to know. To consider that, one must describe the notions and the behaviors, in terms proper to *Didactique*. One must then study the conditions of the processes of formation of knowledge in the students (shown by the expected behavior), in particular those which can be verified or realized by the teacher (teacher's strategies.)

Brousseau thus included in the field of *Didactique* not only theoretical questions but also concrete, practical ones. Experimentation was to take place in the classroom, with analyses designed for the purpose and fully carried out. On the other hand, the objective of the analysis was not the improvement of a particular process or the smoothing of a particular lesson sequence, but rather the building of an entire theory that would serve as a foundation upon which to build didactical choices and decisions. Materials collected in the course of a teaching sequence were to be used to study the conditions governing the behavior of the students and the teacher, and the entire teaching project and thence to deduce the model of interaction that might best account for what had transpired. The methodology (and this is a recurrent theme) was to be that of the experimental sciences.

Thus, by the mid-1970s, Brousseau's vision of *Didactique* as a field of scientific research into the teaching and learning of mathematics was articulate, explicit, and very much a part of the discussion at the national level. He was, in fact, quite widely known by then, in an intriguingly non-intersecting collection of circles. Thanks to the breadth of his interests and his energy in following up on them, he wound up being asked to speak or write articles or be on masters' degree committees by people in a number of different fields, each of whom assumed that his specialty was close to theirs (from fields as divergent as statistics, psychology, pedagogy, otolaryngology, and the study of mathematical automata.) Geographically, though, he remained firmly based in Bordeaux. He appreciated the support and level of cooperation of his community there, and was not convinced that it could be matched. In fact, he turned down an offer of a more prestigious position in Paris so as to remain in that community. The one change that his status did undergo was that in 1970 he became an Assistant in the mathematics department at the University of Bordeaux (which was a cut below elementary school teaching on the pay scale.)

Meanwhile, while spending a formidable amount of energy arranging for Bordeaux's *IREM* and *COREM*, Brousseau had also been developing his own theory of teaching and learning. He built it up from his own teaching experience over a number of years, supported by his and Nadine's experiments and those of the group of teachers he had been working with, and in 1970 he decided it was time to go public with it. He was invited to be one of a pair of speakers at a meeting of the Association of Teachers of Mathematics addressing the issue of "The Process of Mathematization of Reality." His fellow speaker discussed the teaching of probability. Then the audience turned to him, expecting to hear about the teaching of statistics. Instead they received a full-blown description of the Theory of Situations. They came out very much impressed, if a trifle bewildered—Brousseau reports that the standard reaction was, "That was very interesting, but we didn't understand any of it!"

His initial description of the theory makes it clear both why his colleagues felt that his ideas merited study and why they felt the need for clarification. It included the following:

> The structure of a set is defined by the relations and operations which connect its elements. ... Let us consider a Situation, that is, a certain organization of objects or of persons having certain relationships among them. To describe this Situation, it is sometimes convenient to choose a structure and to establish, between certain of its elements or relations and the objects or relations of the Situation some correspondence of signifier to signified. ... It can happen that the consequences in the structure signify concretely the consequences in the Situation. In that case, the structure will permit valid predictions. A structure envisaged in a certain Situation as a means of predicting and explaining, with the project of extending the parts giving concrete significance, constitutes a model of the Situation. It can happen that not all the relationships in a mathematical model can be given a concrete significance in the Situation described: these relations are said to be non-pertinent. It can also happen that the consequences in the Situation do not match up to those predicted by the model. In that case it should be rejected or modified. (Brousseau 1972)

This is not a description which causes the theory to leap instantly into clarity, but it does establish the objective of creating a mathematical model of the teaching and learning of mathematical knowledge. During the next few years a number of elements became clearer and more explicit. The search was for a series of Situations characterizing a particular piece of knowledge. Very early on, the Situations and the possible actions of the students began to be described in the vocabulary of the theory of games.

A colloquium talk that Brousseau gave in 1975 provides a good deal of clarification of the initial description. In it he characterized *Didactique* as follows:

> The underlying hypothesis of *Didactique* is that a learning process can be characterized by a sequence of reproducible Situations that lead to the students' learning of a particular piece of knowledge, or more concretely to a set of modifications of the students' behaviors which characterize the acquisition of that piece of knowledge.
>
> Each such Situation brings together three components:
>
> some mathematical knowledge in its full richness—not merely a statement of a mathematical concept, but its meaning, its uses, its connections to prior knowledge, the context in which is it likely to be encountered, the language commonly used to express it,
>
> some subjects—students regarded from the point of view of their initial states and their manner of evolving, responding and learning,
>
> and some "didactical wherewithals", which later came to be called the "*milieu*"—the teacher, the materials and the learning strategies chosen.
>
> The objective of research in *Didactique* is to study the ways in which these situations evolve, to characterize them in terms of the didactical conditions of learning and of the informational content of specific concepts and to find the "laws" governing their evolution.

Some Specific Ideas and Situations

A closer look at one of the earliest Situations and how it was used seems worthwhile at this stage. Let us look at the Race to Twenty. A description and discussion of it by Brousseau himself constitute the introductory chapter of the Theory of Didactical

Situations in Mathematics (Brousseau 1997). What follows will be a little more descriptive and a little less technical. The didactical Situation itself is based on one of the well known NIM games: the first player writes either one or two, then on successive turns players alternate writing numbers, each of which must be either one or two greater than the one previously written. The objective is to be the one who writes "20". What turns this from an activity into a Situation is a very specific structure for its use:

1. The teacher states the rules concisely, then invites a student to the board to start playing a round of the game with her.
2. Halfway through the game, she calls on another student to take over for her, so that she herself neither wins nor loses.
3. Students are then set up in pairs to play against each other and observe whatever they can about the game.
4. After 10 min, the class is assembled into two groups. There follows an alternation of group planning time and rounds of the game in which randomly selected individuals from each group play against each other at the board. While they are playing, their group cannot coach them.
5. After six or eight rounds, the teacher launches a "game of discoveries", in which the groups alternate making propositions about the Race to Twenty, such as "If you write 17, you'll win". The other group must either accept a proposition or attack it. The latter case (obviously) can generate interesting discussions.

What are the salient characteristics of this Situation? Well, for a start, as with the example of the pots of paint, for the right level of student it can be highly engaging. Furthermore, the teacher's job is to unobtrusively keep everything running, but never, even at the fifth stage, to be the source of information about what is correct in the mathematical thinking, with the possible exception of maintaining standards for the articulation of the arguments. Less apparent is its position in a curriculum or program, but it certainly has one. Brousseau came up with the basic idea sometime in the 1960s in addressing the issue of students struggling with division. An internationally familiar—and internationally unacceptable—phenomenon is that of a student who has been defeated by a concept going back to the beginning and retracing the same steps to the same defeat. Brousseau's idea was to put students in a position where they needed division and could re-invent it and become secure with it before they realized that it was, in fact, the concept that had given them so much trouble. So he started them with the Race to Twenty. What they discover in the original form is that if they play the sequence 2,5,8,11,14,17, they will win. Then he played with the two variables in the game—the goal number and the largest number that can be added onto an opponent's number. As long as the largest number that can be played is 2, the winning sequence can always be determined by subtracting 3 repeatedly from the goal number. On the other hand, if the players are permitted, for instance, to add 1,2,3,4 or 5, then the number to be subtracted repeatedly is 6. In general, whatever the largest number is that can be added to the opponent's play, the next number larger is the one which must be subtracted. The requirement for winning is that the first player subtract the number in question often enough to get down to a number within the permissible writing range, and start with that. Thus, once the game is well established, the players know that victory logically depends on the selection of the very first number. So to start a "Race to 23" where the rules permit adding 1, 2 or 3 you start at 23 and count backwards by 4s : 19,15,11,7,3 so you must start with a 3. But counting backwards is increasingly unnatural, so it becomes more economical to start subtracting off 4s (or 7s, or whatever the particular round calls for.) That will always work, but suppose you were playing "Race to 157"? How often do you really want to subtract? Well, maybe it would be more efficient if we were to bunch some of those 4s—for instance bunch 10 of them and you can already subtract off a 40 .... After enough of this, the student is ready to discover, or to be led to observe, that this process of repeated subtraction is really a form of division, and the vitally important number is simply the remainder. Note, though, that phrase "enough of this". That's where the teacher supercedes the theory—no teaching algorithm can replace a teacher's understanding about the individual student in determining when the student has had enough experience at one level to be led into the next level.

That, then, is the point of the Race to Twenty from a teacher's-eye view. From the perspective of the didacticians, however, it has a far broader context. In the early days of the field the Race to Twenty was one of the most heavily used tools for research. By the time Brousseau wrote the article about it that he published in 1978 he could already report that it had been carried out 60 times, which provided an excellent consistency check—teachers with different styles and personalities could use the Situation, and highly similar developments would result. Many specific aspects were studied and statistically verified. One that I find particularly intriguing is that whereas a class can be depended on to discover the importance of the number 17 quite swiftly, the average time for discovering each successive member of the descending sequence is longer, all the way down. Other results about further aspects continued to appear into the 1990s.

Another use that was made heavily of the Race to Twenty in the early years was to illustrate the *Theory of Situations*. The theory has developed greatly since then, and become correspondingly more complex, but this remains an excellent launching pad for describing its most vital elements.

Three Basic Situations

A central idea in the theory is to focus on three Situations (briefly, early on, called dialectics): that of action, of formulation (later sometimes called communication) and of validation. They must not be regarded as discrete categories—in fact, one of Brousseau's formulations (see below) has each one imbedded in its predecessor—but they certainly represent different attributes. In the Race to Twenty, the Situation of **action** presents itself very clearly: once the teacher has given and clarified the rules of the race, students settle in to play it. As they do so, they develop an implicit model of a strategy, in fact a succession of implicit models, some of which render previous models obsolete. For an observer, these models can be seen by, for instance, the frequency with which the number 17 is played, long before a student is even conscious that she is playing it, much less able to explain why. The Situation of action lays the essential foundation for the explicit models and formulations which follow. A similarity worth noting is that one of the features of teaching with "manipulatives" that is most uniformly agreed to (at least in theory) is that an essential first step is to allow enough time of free play for the students to form a model of how these objects work. A specific requirement in the case of *Didactique* is that the Situation of action provide feedback to the student on which to base, and against which to test, his models. In this case, the feedback consists simply of the winning or losing of the race.

In the Race to Twenty, once the individual playing phase has ended, a new phase begins. Students are in two groups, and periods of group discussion alternate with games played at the board by randomly chosen individual students, one from each group. While they play, the rest of the group must remain silent. During the play at the board, the Situation is again one of action for the players. During the discussion phases, on the other hand, since no one knows who will be chosen as the team's next representative, the only way to win is to communicate one's ideas to the whole group, which requires formulating them and making them explicit. This therefore represents the second type of Situation, that of **formulation**. In general, formulation occurs in Situations where the student has a certain amount of information, but either needs more information than she can come up with on her own or does not have the means of taking action on her own, and in order to proceed must communicate with other members of the class. If the groups become argumentative, the next Situation may develop while the group planning sessions are going on. In any case, it will do so in the following one. This is the Situation of **validation**. Underlying the importance of Situations of validation is the fact that part of mathematics is the development and conveying of conviction. A mathematician presenting or publishing a theorem can't just be pretty sure she is right. She must be so certain and so clear about it that she can convince others of her correctness. Ultimately that convincing takes the form of formal proof, but the ability to produce such a proof develops very gradually. Situations of validation support that development and help the students progress from the use of decibels and repetition (the tactics natural to young children) toward the use of logic, language and clear thinking.

In this particular sequence, the validation itself is made into a game whose rules are somewhat loosely constructed so that the teacher can use her judgment to decide when and how strictly to apply them. Groups alternate making "declarations", and score one point for a declaration simply accepted, or two for a successful disproof of a declaration by the opposing team or a successful defense of a declaration which the opponents have challenged. The teacher needs to exercise a certain amount of control to keep the social dynamic natural to children from overwhelming the mathematical one—that is, to make sure that decisions are made on the basis of logical clarity, not on who talks loudest or has a position of authority within the class.

This presentation of Situations is now the most common one. It may be helpful, though, to consider an earlier presentation of them. In his article on the Race to Twenty, Brousseau finished his discussion of the three Situations with a statement which he elaborated further in 1978, to the effect that the second Situation is imbedded in the first, and the third in the second. All three can be regarded belonging to a Situation of action. In the course of interacting with the initial *milieu*, the student will necessarily be formulating some ideas about those interactions. If the circumstances require it, that formulation becomes explicit, in which case that part of the action Situation becomes a Situation of formulation. The formulation could simply be a useful tool, but if in addition it requires defending, then part of the formulation Situation becomes a Situation of validation.

Non-situations

With the preceding Situation freshly in mind, this seems a good point to point out a distinction. The Race to Twenty (or to Seventeen, or whatever) appears in a number of collections of games for use in the classrooms or in family math nights or the like. It works very nicely for them, in fact, but as an activity, not a Situation. One contrast is the structure: note how carefully the Situation is set up to encourage autonomous learning on the part of the students, and how explicitly the order and timing of the different portions of it are arranged. Another is the goal: this is the introductory portion of a longer Situation designed specifically to induce the students to create (or re-create) for themselves the concept of long division. A third is the research aspect, in fact, two research aspects. For one thing, no Situation is fully, seriously accepted until it has been "scientifically" tested—that is, until it has been studied in the abstract, with predictions about the outcomes of different phases of it, run under close observation, and analyzed with respect to the predictions. It is also true that many Situations have been used as tools to further research into how children learn, and in particular how children acquire and manifest an understanding of a given concept.

An activity can serve many functions, from livening up a classroom to demonstrating to parents what kinds of thinking are involved in current mathematical education. It can also serve as a tool in the construction of a Situation. What it does not do is stand on its own as a Situation.

This seems a good point for another clarification, this one not of concept, but of orthography. A few years ago I had a rather heated discussion with a colleague who had read an early version of this manuscript. Once the air cleared, it turned out that she thought I was making grandiose claims about ordinary, everyday situations when in fact I was talking about didactical ones. To prevent this misapprehension from disturbing anyone else, I decided to use a capital S when and only when I am referring to one of the elements of *Didactique*. It looks a little bizarre, as you have no doubt observed, but it does serve a function!

Didactical and A-Didactical Situations

Another use of the word Situation needs to be brought up at this point. It describes a dichotomy that is useful in the analysis of teaching with Situations, even though the separation point is not necessarily unambiguous. A Situation is a-didactical if the teacher's specific intentions are successfully hidden from the students and the student can function without the teacher's intervention. This doesn't mean that the students think the teacher is in the room simply as an entertainer, but that they are not conscious of the specific intent to have them learn some specific concept. For instance, while they play the Race to Twenty, the students are probably aware at some level that the teacher has something in mind beyond the game itself (especially since it is played during math period). On the other hand, the fact that this is a step toward learning or reinforcing the concept of division is totally out of their line of sight, and remains so for a considerable period. Finding Situations of this nature was central to the early stages of the field. Only later did the researchers realize the importance of the other kind—the didactical Situations.

Situation of Institutionalization

With the structure of Situations described above, Brousseau and the others who had joined his research efforts began applying, testing and elaborating the Theory of Situations. Much of the experimentation was done at the COREM—the École Michelet which had been set up for observation. Things were proceeding smoothly with one of the sequences when, at a point when the research team said to go on to another set-up, the teachers, led by Nadine Brousseau, said "No, not yet." After an initial reaction of consternation, the researchers, who had a great deal of respect for the teachers, settled down to analyze this firm resistance. Thus was born a fourth Situation, that of **institutionalization**.

In one of his papers, Brousseau gives a description of the process. As he describes it the researchers were feeling just a trifle smug about having envisaged all types of Situations. They were perplexed at the requests of the teachers for more space—a pause before going on, because they needed to review what was going on, or to do something about some students who were lost and needed to be gathered back in. Eventually the researchers stopped and took stock of the overall state of affairs. Their whole focus had been on a-didactical Situations. They had constructed all sorts of a-didactical Situations, in which the teacher's entire function was to arrange the set-up and stand back while the students constructed their knowledge. What they hadn't realized—but did as a result of this impasse—was the teacher's role in making all of this cohere. The teacher is needed to record what the students do, to keep them conscious of what they have developed and how it is related to what they did before, to give a status to various elements so that they can be used for reference later on, and to build the bridges between what the children have produced and what is known and accepted in the culture at large. If necessary, the outside connection can include the teaching of a more conventional notation or the labeling of terms in a way that is consistent with general practice. This is an essential element in making the knowledge the students have constructed available and useful—a fact which the researchers eventually grasped, thoroughly appreciated, and accorded the title of *institutionalization*.

Classical teaching methods (internationally!) are based on institutionalization alone, without the creation of meaning: say what is to be learned, explain it and test for it. The researchers had been obsessed by a-didactical Situations because they are precisely what is traditionally totally lacking.

The four types of Situation, action, formulation, validation and institutionalization, lie at the heart of the structure of the Theory of Didactical Situations. Other major concepts contributed to the elaboration of the theory and were in turn enriched by the research it engendered. Two notable ones are **epistemological obstacles** and the **didactical contract**.

Epistemological, Ontogenic and Didactical Obstacles

One common misapprehension, particularly among those who have not actually taught, is that if teaching is done right, no misapprehensions need ever develop. Brousseau describes this as a belief in an enchanted learning in which the volume of learning swells in an empty space, at all times correct, structured, usable and transferable. This is, alas, an illusion. A more accurate vision must take a closer look at the meaning of learning.

An underlying hypothesis of *Didactique* is that a mathematical notion is only understood if it has some function, that is, if it successfully solves some problem. For the learner, the meaning of the notion is given by the problems it solves. Learning it means being able to put it to work in some domain and have it dependably function there. The snag is that that domain rarely starts off completely general. The learner will use the notion within its initial domain, and in time it will develop a value and consistency and meaning that may make its generalization or modification progressively more difficult. For later acquisitions it becomes both a support (because the learner already knows something) and an obstacle (because part of what the learner knows is false in a wider context.)

There are many examples, but the most accessible of all is number: one of the earliest notions that children are specifically taught and loudly praised for is counting. This produces a initial understanding of number which includes the notion that after every number there is a well defined next number. Some time later, a child learns to add and multiply, and internalizes the idea that multiplication makes a number bigger. Both of those are essential to a child's understanding—so much so that nobody would say that children with those two beliefs are badly taught. On the other hand, both of those ideas become false when the notion of number is expanded to include fractions. The ideas are not mistakes, but they do constitute obstacles.

This gives rise to Brousseau's formulation that new learning occurs both from and against old learning—against in the sense that the learner must struggle to sort out and eliminate the elements of the old learning that have become false in the new setting. The title of **epistemological obstacle** is given to the part of the old learning that occasions the struggle — the part that has become false in the interim. *Didactique* is thus presented with the challenge of creating Situations in which the old learning can be used (initially), be seen to need replacing, and support the production of a new learning meeting the new needs. The characteristic status of an error of **epistemological origin** is that to some extent it remains part of the fabric of the concept, so that part (though definitely not all) of the learning process still needs to proceed through it.

Another type of obstacle is one of **ontogenic origin**. Those are the ones that occur because of limitations (neurophysical, among others) of the subject at a particular moment of her development: she develops the knowledge appropriate to her abilities and goals at a specific age. One good illustration is an extremely common number difficulty in young children. A child of five, or thereabouts, can subitise (distinguish between sets containing different numbers of objects without recourse to any explicit counting process) sets containing up to 6 objects. Subitising becomes unreliable and difficult for 7, and breaks down entirely after 8. Attempts to teach the numbers 6, 7 and 8 directly cause much difficulty and distress. On the other hand, if the child is asked to compare sets with 10 or 15 objects, doing it by straight perception is so obviously disadvantageous that the child will generally opt straightaway for creating new procedures requiring breaking the sets apart and using counting skills. The new procedures involve a kind of implicit addition, and thus provide a foundation for the later learning of addition.

Another type of error is one of **didactical origin.** Those are the ones that result from choices made by the school system itself. Brousseau cites several that arise from the choice of the French educational system to teach decimal numbers very early as a computational device and to relate them to real life by including many examples involving measuring. In that context, and with that level of sophistication, students pick up a very firm message that things beyond the hundredths place are just decorative and can perfectly well be ignored. This undermines efforts to teach approximation (so π remains forever 3.14). It also encourages the belief that every number has a sequel. More subtly, and therefore more seriously, it is a major obstacle to the learning of limits, because it makes it very difficult to believe that anything important can happen after the terms get small.

How then are these obstacles to be used? They are essential in the design of Situations. If a concept is one around which these obstacles lurk, then any Situation designed to induce students to construct that concept had better take the obstacles into account.

The Didactical Contract

Of all of Brousseau's ideas and conceptions, the one that has caused the most frustration in the world of English-speaking mathematics educators has been the didactical contract. Just enough has been known for many to recognize it as significant and interesting, but no more—a highly tantalizing state of affairs. Personal experience has now revealed to me how easy it is for any simplified description to produce misunderstandings. This, together with the fact that many aspects of the Theory of Situations itself are involved with the derivation and understanding of the didactical contract, has prompted me to include rather an extensive discussion, which will then be filled in yet further in Chap. 4.[7]

First, however, a brief description of its origins: in the early 1980s, Brousseau and a group of colleagues from several fields, including psychology, undertook a study of a group of students all of whom were failing in mathematics, but only mathematics. One of them was an 8½-year old boy named Gaël, bright, cooperative, acquiescent and astoundingly non-functional in mathematics. After checking that his developmental level was normal, they began to inspect other aspects of his situation. Eventually they realized that he had quietly negotiated a "contract" of which no one (including himself) was aware: his part of it was never to risk failure by actually trying to understand the mathematics, but always to maintain his equanimity and his acquiescence; the adult's part was to refrain from chastising him or rendering him uncomfortable, and to supply the answers to the questions she herself had posed. A very pleasant arrangement all round, aside from the fact that Gaël was learning absolutely nothing. Having formulated this theory, the group verified it by bending the contract slightly and observing the results, then figured a way to break the contract altogether and set up a new one. Gaël returned to the classroom with a whole new concept of himself as a learner of mathematics, and was able to maintain the change and keep up with his class thereafter.

Gaël's development is described in detail in *The Case of Gaël,* which is summarized in Chap. 4.

Once Gaël was back in the classroom, Brousseau and others got busy applying what they learned from working with him to what they had already learned in the Theory of Situations. It fitted in so well that a good description involves a re-description of many aspects of the theory. First, a capsule version of the relationship between the mathematics of the mathematician and the mathematics of the classroom:

Contrary to common public image, mathematical results do not leap full-blown into the mind of a mathematician. They are produced in some context, generally with initial errors and gaps that the mathematician, with time and effort, sorts out. Then, before the results are made public, in order to maximize their usefulness the mathematician works to de-contextualize, de-personalize and de-temporalize them.

The work of the teacher is the reverse: to make a piece of existing mathematics accessible, she must re-contextualize and re-personalize it so that it will have meaning for the student. If this goes really well, the student will have a deep understanding of the concept, but not necessarily any realization that it has a general meaning. So then the teacher must assist the student in re-de-contextualizing and re-de-personalizing the result so as to grasp its universal character and its status as cultural knowledge.

---

[7]The discussions here and in Chap. 4 are based on Brousseau, G. and Warfield, V., The Case of Gaël, The Journal of Mathematical Behavior, Vol. 18, no. 1 (1999) pp. 7–52.

That means the teacher has a two-stage role: first to make the knowledge alive and personal for the student by setting up a familiar situation for the student for which the development of the knowledge in question will be a reasonable response, and then to give that "reasonable response" a form and status that connect it with outside culture.

Clearly, given time and energy constraints, a teacher is subject to the temptation to take a shortcut and simply present the knowledge directly as a cultural fact to be accepted and filed somewhere in the student's brain, leaving the student to make sense of it on his own — or not.

If we agree that this shortcut is not acceptable, then the teacher faces a complex set of demands. The teacher must find a Situation in which to place the student. The Situation must not be one in which the answer to the question posed is immediately apparent (in which case little or nothing is being learned) or in which it is so obscure that the student is defeated and won't undertake the search. The student needs to be able to produce some solution using prior knowledge and then to see the need to modify it, and be motivated to keep on modifying it until he arrives at a successful solution. The more it needs modifying, the more engaging the Situation needs to be to keep the student from giving up.

In other words, the teacher's work is to propose to the student a learning Situation such that the student produces the piece of knowledge that is to be learned as his own personal answer, and uses it or modifies it to satisfy the constraints of the *milieu*, and not the expectations of the teacher. There's a big difference between finding something because the teacher wants it and finding something because the circumstances are such that it is really needed. The teacher's intentions and expectations can never become completely invisible, but the greater the extent to which the student can forget them, the more the learning becomes the student's own.

That is not an easy task. Even in life, people have a tendency to regard things that happen as being designed to teach them a lesson. In the classroom that tendency has a good, solid foundation in reality. For a child to read a Situation as being independent of the teacher's will, the Situation must be what Brousseau calls "an intentional cognitive epistemological construction", that is, it must be designed with an eye constantly on its capacity to be turned over to the student and have the desired consequences. With such a construction, the solution of the problem can become the responsibility of the student. Even then it doesn't happen automatically. Simply communicating a problem to a student doesn't guarantee that the student will feel responsible for solving it. And if the student does accept the responsibility, there is still no guarantee that he will be able to see it as a universal problem rather than a personal one.

The activity of the teacher in attempting to induce the student to take on responsibility for a Situation is given the title of "devolution".[8] This devolution is essential and central but there is absolutely no way to guarantee that it will succeed. This

---

[8] Devolution was an act by which the king, by divine right, gave up power in order to confer it on a Chamber. "Devolution" signifies "It is no longer I who wills, it is you who must will, but I am giving you this right because you cannot take it yourself".

leads to the next level of consideration: what happens if the student refuses to take on the problem, or else can't solve it? Then the teacher has to help, and sometimes even has to defend herself for having given such a difficult problem.

It is the negotiating of this phase which produces, or at least brings into play, a complex set of relationships of obligations between teacher and student. Sometimes explicitly, but more often implicitly, a determination is reached about what each has the responsibility for managing. The resulting system of reciprocal obligations resembles a contract. The part of that contract that is specific to the target mathematical knowledge is called the *didactical contract*. It is not possible to give details of these reciprocal responsibilities—in fact, it is the study of the breaking of the contract that is most revealing.

Let us have a look at the underlying situation. It is assumed that the teacher will create conditions that invite the appropriation of knowledge and will give recognition to that appropriation when it occurs. It is assumed that the student will respond correctly to those conditions. It is clear that the didactical relationship must continue at all costs. The teacher therefore must assume that earlier learnings and the new conditions make it possible for the student to achieve the new learning. If the learning does not happen, both the student and the teacher are blamed.

It should be recognized that this interplay of obligations isn't an actual contract. For a start, insofar as it concerns the result of a teaching action it can't be made completely explicit. There are no known, recognized ways of producing the construction of new knowledge or of guaranteeing that a student will appropriate the target knowledge. On the other hand, if the contract involved only the rules of behavior of the teacher and student, scrupulous obedience to it would produce a classroom where behavior took consistent precedence over substance—decidedly a failure of the didactical relationship.

The teacher must see to it that the student really can learn the target knowledge, and assume responsibility for the results. That is actually an impossible hypothesis, but it is absolutely necessary if the teacher is to be allowed to engage the student's responsibility. Similarly, the student must accept responsibility for solving problems whose solutions she has not been taught, and assume that it is always possible to do so, even though she can't see her way through from the outset.

Clearly, these responsibilities, even though accepted, can't be described in detail when they are taken on. Nor can advance notice be given of what constitutes breaking the contract, or what the consequences will be. Nonetheless, when process does break down, everyone behaves exactly as if some implicit contract had been broken. The student is surprised and annoyed because the teacher has asked the impossible, the teacher is surprised and annoyed because the student isn't carrying out what looked to be a reasonable project. There is a revolt, a negotiation and a search for a new contract that depends on the new state of knowledge, acquired or desired.

It is at this phase that observation and study can be made. Hence in some sense the issue about which *Didactique* can build and study a theory is not the didactical contract itself, but rather the process of finding and sustaining such a contract.

Examples illustrating devolution can be found in *Theory of Didactical Situations in Mathematics,* but for reasons which should by now be clear, a recommended didactical contract, in bullet form and with sub clauses, is not an available option. Underlining this impossibility further is the issue Brousseau calls the **paradox of the devolution of Situations**:

> The teacher must see to it that the student solves the problems set for him, so that both of them can assess whether he has accomplished his task.
>
> On the other hand, if the student produces an answer to the problem without having grappled with it and made the choices required for really understanding it, then the evidence given by his having an answer is misleading. If, for instance, the teacher has been induced to tell the student how the problem can be solved, or to give some trick for coming up with the answer itself, then the student has had no opportunity to construct the knowledge in question. So the more a teacher gives in to the demands of a student (especially one who is struggling) and tells him precisely what he must do, the more she risks losing her chance of obtaining the learning she is aiming at.[9]

This means that the didactical contract presents the teacher with a paradoxical injunction: her aim is to lead the student to learn and understand some concept, and her indication that that concept has been learned will be some set of behaviors on the part of the student, but anything that she does that aims directly at producing that set of behaviors deprives the student of the opportunity to learn the concept itself.

By the same token, the student is also presented with a paradoxical injunction: if he accepts that the contract requires the teacher to teach him everything, he doesn't establish anything for himself, so he doesn't learn any mathematics. On the other hand, if he refuses to accept any information from the teacher, the didactical relationship is broken and he can make no progress at all. In order to learn, he must accept the didactical relationship but consider it temporary and do his best to reject it.

In the Guise of a Conclusion

The final section of articles on *Didactique* is frequently entitled *"En guise de conclusion"*. The correct translation is "By way of a conclusion", but for my purposes I prefer the more literal one above. I shall close this chapter with a lengthy quotation from an early article by Brousseau, presented in its full original glory, uncushioned by interpretation or explanation. To me it illustrates the breadth and scope of his thinking, and the vigor (or perhaps the word is glee) with which he responds to the fact that delving into the questions, the concepts, the convolutions — in short, the entire theory that he has launched — inevitably produces at least as many challenges as it does solutions. To me, it demonstrates why the study of *Didactique* is exciting, productive, worthwhile — and yes, at times daunting!

> *There is another thing that we took a long time to notice. Initially, we implicitly thought that learning situations were almost the only means by which knowledge is passed on to students. This idea arises from a rather suspect epistemological conception, as an empiricist*

---

[9] Personal communication from Brousseau to Warfield, 2004.

idea of the construction of knowledge: the student, placed in a well chosen situation, should, on contact with a certain type of reality, construct for himself knowledge identical to the human knowledge of his time (!) This reality can be a material reality in a situation of action or a social reality in a situation of communication or of proof. We know very well that it is the teacher who has chosen the situations because she was aiming at a certain piece of knowledge—but could it coincide with the "common" meaning? The student had "constructed a meaning" but was it institutionalizable? One could proceed to an institutionalization of pieces of knowledge, but not to that of meaning. Meaning put into a situation cannot be recovered by the student; if the teacher is changed, the new teacher does not know what has been done before. If one wants to review what has been done, one must really have the relevant concepts for this purpose, they must be universal and they must be able to be mobilized along with the others.

Meaning ought also to be institutionalizable to some degree. Let us see how. This is the most difficult part of the teacher's role; to give meaning to pieces of knowledge, and, above all, to recognize this meaning. There is no canonical definition of meaning. For example, there are social reasons why teachers have stuck to the teaching of the division algorithm. All reforms try to operate on understanding and meaning, but generally they are not successful and the objective of the reform appears contradictory to the teaching of algorithms. Teachers have cut down to what is negotiable, that is to say, formal and dogmatic learning of knowledge, because the moment when this has been done can be identified by society.

There is the idea that knowledge can be taught but that understanding is in the student's hands. One can teach the algorithm and "good teachers" then try to give it meaning. This difference between form and meaning results in difficulties not only in conceiving of a technique for teaching meaning, but also of a didactical contract for this purpose. Otherwise stated, one will not be able to ask teachers to use a Situation of action, of formulation, of proof, if one cannot find a way of allowing them to negotiate the didactical contract connected with this activity; that is to say, if we cannot negotiate this teaching activity in usable terms.

In geometry, for example, let us imagine that we would like to encourage the student's mastery of spatial relations. It is difficult to negotiate this objective, except in very small classes, because it doesn't exist as an object of knowledge. It is confused with the teaching of geometry which has nothing to do with it. It is not true that geometry teaches spatial relations.

There are a certain number of mathematical concepts which hold no interest for mathematicians, but which do for Didactique, and which, because of this, have neither cultural nor social status. For example, the enumerating of a set isn't a mathematically important concept, but it is an important concept for teaching. Has Didactique the right to introduce concepts which are necessary for itself into the field of mathematics? This is a subject which we shall have to discuss with the mathematical community and with others.

Negotiation, by teachers, of the teaching of understanding and meaning poses a true didactical problem, a technical and theoretical problem for the didactical contract. How should we define and negotiate the object of the activity with the public, the teacher, the student, and other teachers?

The reader knows very well that there are several types of division, but we have only one word. Division by whole numbers and division by decimals refer, in fact, to different conceptions; this causes a lot of problems. Teachers don't have the possibility of possessing an object which could be called "the meaning of division", which they could say they were working on.

We are attempting to provide a didactical model of meaning negotiable between the teacher and the student, which would allow the student to work on the meaning of division with a vocabulary and concepts that are acceptable and that truly develop her knowing, and which would consist of Situations in which she does divisions. This meaning consists of classifications, tools, a terminology. But there is danger in work of this sort; the development of a species of pseudo-knowledge or a ridiculous, useless "mis-knowing".

It must not be thought that Didactique consists only of presenting as discoveries the things that young children do. Problems must be solved by means of theoretical knowledge and by technical means. It is necessary to propose something in order to have an effect on certain teaching phenomena; but it is necessary, first, to identify them and to explain them. The work of managing the meaning of the didactical contract, relative to meaning by the teacher or between teachers of different levels, is a delicate theoretical problem and one of the principal stakes of Didactique. Today, teachers of different levels give conclusions which tend to produce a collapse of lower-level activities onto more formal activities because they cannot negotiate anything else.

The reprise by a teacher of all the non-institutionalized old knowing is very difficult. In order to produce new knowledge, she can to some extent use pieces of knowledge that she has tried to introduce herself; this is not very easy. But when it is someone else who has introduced these pieces of knowledge, and if they have seldom been used, the problems become almost insurmountable. The only way to manage is to ask teachers of lower classes to teach, in a rather formal manner, knowledge that the teacher of higher classes identifies, which can serve her at the explicit level in order to construct what she wants to teach.

We know nothing much about interactions between didactical activities; how are they managed over time? We must therefore evolve our conception of the construction of meaning.

Chapter 2
Two Illustrative Examples

In the course of the past 30 years many sequences of Situations have been invented and used and experimented with. The level of student aimed for ranges from pre-school to university, with the highest density at the fourth to eighth grade level. Of these, two are completely or partially available in English. A sequence on introducing statistics, with commentary by Brousseau, was published by Brousseau and Warfield in the Journal of Mathematical Behavior.[1] That sequence serves as an introduction to Fundamental Situations, and will be discussed in detail in Chap. 3. The other sequence is far longer and has been taught many times. It provides a complete introduction to rational and decimal numbers as required by "obligatory scholarship" (in the National Programme).[2] A detailed presentation of that sequence, broken into several sections to keep the size manageable, has been published in the Journal of Mathematical Behavior.[3] Two particularly intriguing Situations appear early in the sequence and, as with the pots of paint and the Race to Twenty, provide in themselves a nice illustration of the nature and content of a Situation. This chapter contains both, with a sketch of the connecting material between them.

Before presenting the material, I need to give a little explanation of its background and format. In the early 1970s, the Theory of Situations was well developed in Brousseau's head, but he felt a strong need to prove by experimentation that it could succeed in function as well as in theory. He had the new COREM—the Michelet School—ready and waiting for such an experiment, and the major decision was what concept or sequence of concepts to work with. One that he chose was rational and decimal numbers. He made the choice for both mathematical and technical reasons—the latter being the fact that his presentation could be substituted for part of the Programme without horrendous distortion of the normal sequence. After intensive study, he came up with a sequence of Situations that were then

---

[1] Brousseau, G., Brousseau, N. and Warfield, 2001.

[2] *Scolarité obligatoire.*

[3] Brousseau, G., Brousseau, N. and Warfield, 2004, 2007, 2008, 2009.

V.M. Warfield, *Invitation to Didactique*, SpringerBriefs in Education 30, DOI 10.1007/978-1-4614-8199-7_2, © The Author(s) 2014

taught by his wife, Nadine, and the other fifth grade teacher at the school. The following year they were taught again with some minor modifications, and thereafter they became part of the standard curriculum at that school. In the course of the next 8 years, Nadine continued to teach the sequence, keeping careful notes as she did so. Eventually the two Brousseaus were persuaded to combine their expertises and write a manual, which was produced in mimeographed form by the IREM of Bordeaux.[4] It is a remarkable and rich document, in part because of the instructions it presents for carrying out the sequence of Situations, and in part because of the commentary that accompanies the instructions. What the reader needs to realize is that the frequent remarks to the effect that "at this point the students have grasped X, but not Y", or "some of the students have understood this, but most will need reinforcement" are not theoretical comments based on the image Guy Brousseau had in his head as he invented the Situations, but information backed by 10 years of careful note-taking by Nadine Brousseau as she presented the Situations to a whole succession of fifth grade classes. They need to be read with that understanding.

Part of Brousseau's preparation for creating the Situations was a study of the historical development of rational and decimal numbers and of the course of development of the teaching of them. Brousseau developed a firm commitment to the idea that a major source of students' conceptual difficulties is the failure to sort out the different ways in which fractions function, in particular as measures and as linear mappings. Correspondingly, he put together an initial Situation that sets students up to create rational numbers strictly as a method of measurement.

A Didactical Situation: The Thickness of a Sheet of Paper

Preparation of the Materials and the Setting

The teacher has arranged the following:

- On a table in front of the children, five piles of about 200 sheets each of paper of the same size and color but of different thicknesses (for instance, photocopy paper and card stock), placed in a random order. Different piles can have different numbers of sheets. Some of the differences in thickness cannot be perceived merely by touch. The teacher does not need to know the thicknesses in advance; finding the correct measurement is not the issue;
- In the back of the classroom, on another table, five other piles of about 200 sheets of paper of the same types, in a different order, which will be used in Phase 2;
- Ten plastic rulers (two for each group of four or five children);
- A curtain or a screen that can be used to divide the classroom into two sections.

---

[4] Brousseau, Nadine et Guy, *RATIONNELS et DÉCIMAUX dans la scolarité obligatoire*, Document pour les enseignants et pour les formateurs, I.R.E.M. de Bordeaux, Université de Bordeaux 1, 1987.

First Phase: Search for a Code (About 20–25 min)

The teacher divides the children into groups of four or five.

Presentation of the Situation

Instructions: "Look at the sheets of paper that I have placed in these piles labeled A, B, C, D and E. In each pile, all the sheets have the same thickness, but from one pile to the other, perhaps the thickness is not the same. Can you feel these differences?"

Some sheets from each pile are passed around the class—the children touch and compare them. "What do they do in business to compare the different qualities of sheets of paper?" (Weigh them.)

Objective

"You are going to try to invent another way of denoting and recognizing these different types of paper, and of distinguishing them by their thickness alone. You are in competing groups. Each group is going to think of a way of denoting the thickness of the sheets. As soon as you have found a way, you will try it out in a communication game. You can try things out with the sheets and with these rulers."

Development and Remarks

Nearly all the children try to measure the thickness of a single sheet so as to obtain the required result immediately.

They make remarks like: "It is so thin that one sheet doesn't have a thickness" or "It is very much smaller than 1 mm" or "You can't measure one sheet".

At this stage, there is often a phase of disarray or even discouragement among the students. Then they ask the teacher whether they can take several sheets. Once that idea begins to circulate, they soon try measuring five sheets, ten sheets, or more until they have obtained a thickness sufficiently large to be measured with their rulers. Then they set up systems of designation, such as:

10 sheets 1 mm
60 sheets 7 mm
or
$31 = 2$ mm (The teacher will attempt to get them to realize during the discussion that the latter use of the equality sign is not correct, and suggest an alternative code.)

During this phase, the teacher intervenes as little as she possibly can. She makes remarks only if she sees that in the groups the children are not paying attention to, or have simply forgotten, the instructions. The children can get up, collect sheets, exchange them, etc.

When the majority of the groups have found a system of designation (and all the group members agree with this system or code), or if time has run out, the teacher passes on to the next phase, the communication game.

Second Phase: Communication Game (10–15 min)

Presentation of the Situation

Instructions: "In order to test the code you have just found, you are going to play a communication game. During this game, you will see whether the thickness-code you have established will let you identify the type of sheet designated.

The children in each group will divide into two sub-groups, one of senders and one of receivers.

- All the sender groups will be located on one side of the screen, all the receiver groups on the other.
- Each sender group will choose one of the types of paper on the first table (A, B, C, D or E), which the receiver group cannot see because of the screen. They will send their receiver group a message which must allow them to determine the type of paper chosen. The receivers will use the piles of paper on the second table at the back of the classroom to find the type of paper chosen by the senders.

When the receivers have succeeded in finding the type of paper, they will become the senders. Points will be assigned to groups whose receivers correctly find the type of paper chosen by the senders."

Development and Remarks

The teacher's job is now to

- Pass messages from the senders to the receivers;
- Collect the receivers' replies,
- After deciding whether the reply agrees with the senders' choice, notify the whole group of the success or failure of the message.

All messages are written on a single sheet—which we call the "message form" (see Fig. 2.1)—that passes via the teacher between the senders and the receivers of the same group; this sheet bears the number of the group. In addition, so that the teacher can tell whether the reply is right or wrong, the senders write down on another sheet, which they retain—which we could call the "control form"—the type of paper that they choose for each game.

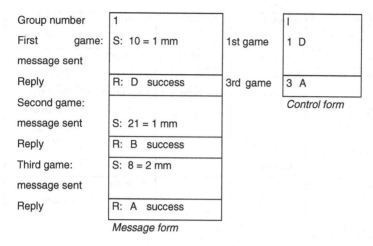

Fig. 2.1 Forms for use in the communication game

Remarks

It is clear that the teacher need not in fact introduce superfluous vocabulary such as "message form" and "control form" nor formal requirements for the presentation of messages — which the children would have to learn to respect. The idea is to give the instructions in such a way that the children grasp the project and make it their own. If this informal presentation results in some group going completely astray from the intended general path, then the teacher intervenes to get them back on track.

If some groups have not been able to construct efficient messages, the teacher has the option of organizing a new phase of discussion, for each group, for the search of a code (with the same instructions as in the first phase). But this situation had never occurred as of the time the manual was written (in 8 identical experiments). The children always succeeded in completing two or three rounds of the game.

During this game, three different tactics are observed among the children:

- Some choose a number of sheets whose thickness they measure;
- Some choose one thickness and count the number of sheets;
- Some look for thickness and number of sheets at random.

Reasonably enough, children tend to choose types of sheet of extreme size, the thinnest or the thickest, so as to facilitate the work of their team mates.

Third Phase: Result of the Games and the Codes (20–25 min)

Presentation of the Situation and Instructions

For this phase, the children come back into their groups of five, as for the first phase of the session. The teacher announces a comparison of the results and prepares on

Type of sheet	Group 1	Group 2	Group 3	Group 4
A	19s, 3mm	10s, 2mm	20s, 4mm	
B	19s, 3mm		4s, 1mm	15s, 2mm
C	19s, 2mm	30s, 2mm	100s 8mm	30s, 3mm 15s, 1mm 20s, 2mm
D	19s, 2mm 12s, 1mm		100s, 9mm	
E			9s, 4mm	13s, 5mm 7s, 3mm

Fig. 2.2 First messages

the blackboard a table showing groups and types of paper, in which she will write the messages exchanged.

Each group in turn provides a "representative", who reads the messages out loud, explains the chosen code, and indicates the result of the game.

The children compare and discuss the different messages. As they are often very different, in the end the teacher requires them to select amongst the codes.

For instance, after discussing the option

10 = 1 mm
TF (for "très fin"—very thin)
60 sheets 7 mm

the whole class might decide to write

10s; 1 mm
60s; 7 mm

This would result in a table like the following one Fig. 2.2:

When all the messages are written down, the children look at the table and spontaneously make remarks such as "That doesn't work" or "Here, this is good", etc.

In general, the resulting remarks can be classified into four categories:

Category 1

If the sheets are of different types, the same number of sheets must correspond to different thicknesses.

Examples from Table 1:

$$
\left.\begin{array}{l}
\text{19s; 3 mm} \rightarrow \text{Type A} \\[6pt]
\text{19s; 3 mm} \rightarrow \text{Type B}
\end{array}\right\} \text{"That doesn't work"}
$$

$$
\left.\begin{array}{l}
\text{19s; 2 mm} \rightarrow \text{Type C} \\[6pt]
\text{19s; 2 mm} \rightarrow \text{Type D}
\end{array}\right\} \text{"That doesn't work"}
$$

Category 2

For the same type of sheet, the same thickness corresponds to the same number of sheets.

Example from Table 1:

$$
\left.\begin{array}{l}
\text{30s; 2 mm} \rightarrow \text{Type C} \\[6pt]
\text{30s; 3 mm} \rightarrow \text{Type C}
\end{array}\right\} \text{"That doesn't work"}
$$

Category 3

If there are twice as many sheets, the thickness is twice as large.

Example from Table 1:

$$
\left.\begin{array}{l}
\text{30s; 3 mm} \rightarrow \text{Type C} \\[6pt]
\text{15s; 1 mm} \rightarrow \text{Type C}
\end{array}\right\} \text{"That doesn't work"}
$$

And the children are encouraged to go on, "We should have found 16s; 2 mm, because:

$$
\begin{array}{ll}
\text{8s,} & \text{1 mm} \\
x2 \downarrow & \downarrow x2 \\
\text{16s,} & \text{2 mm"}
\end{array}
$$

Category 4

Differences in the number of sheets must correspond to equal differences in measurement

Example:

$$
\left.\begin{array}{l}
\text{19s; 3 mm} \\[6pt]
\text{20s; 4 mm}
\end{array}\right\} \text{"That doesn't work because one sheet can't measure 1 mm"}
$$

At the end of the session, the teacher suggests that the children examine the table during the next session, as a class, to verify the measurements by re-measuring and to correct them if necessary.

Remark: The presentation of operations on the numbers using arrows during the search for equivalent pairs is neither formal nor obligatory; it is a familiar manifestation of the use of natural operators which the children have seen in previous classes.

Results (from the data collected in the years of using this material before the manual was written): At the end of this session all the children know how to measure the thickness of a certain number of sheets of paper, to write the corresponding pair, and to reject a type of paper that does not correspond to a pair that they are given. The majority are therefore able to set up a comparison strategy and use it to accept a type of paper as corresponding to a measurement. Some of them have formulated this strategy. Most of the children can analyze a table of measurements and point out incompatibilities by implicit use of the linear model.

Comparison of Thicknesses and Equivalent Pairs

Preparation of Materials and Scene

The material is the same as for Session 1: piles of sheets set out in the same way, rulers.

First Phase (25–30 min)

Presentation of the Situation: Instructions

The teacher asks the children to return to the examination of the table drawn up during the first session.

This observation is first done individually so that the children can spot the most obvious incompatibilities among the measurements. Then the teacher suggests that they fix up the errors that they have seen, line by line (for each type of paper).

Development and Observations

After observation, and on request, the children come up one at a time to the board to indicate the "messages" which seem to them to be incorrect, and possibly to propose a correction.

These corrections are discussed by the whole group of children. If they all agree, the correction is made, otherwise, they suggest concrete verification: they count out the number of sheets indicated by the message and measure them. After collective verification, the new message is adopted and written into the table. (This manipulation is made by groups: two groups verifying one message, two others verifying another message, and so on.) It has often happened that when the same type of paper is measured by different groups of children, these groups do not obtain compatible measurements. This is due to reading errors or to the fact that the children have compressed the sheets to a greater or lesser extent. They are swift to notice this and point it out.

Likewise, for types of paper close in thickness it has happened that the measurements did not permit the recognition of the type of paper in question. It is during this phase that the children are likely to notice that they have more chance of distinguishing between papers of similar thickness by taking a larger number of sheets. Thus, fifteen or twenty sheets of paper of almost the same thickness have measurements so close that the children cannot distinguish between them with precision. At this stage they often suggest measuring the thickness of fifty, eighty, or a hundred sheets.

The following types observations can be made from Fig. 2.2, and have been made in classrooms using this Situation:

For Type A :   we have 10s; 2 mm and 20s; 4 mm. That's fine.
For Type C:

   15 s; 1 mm
   x 2 ↓   ↓ x 2
   30s;   2 mm, so Group 4's other measurements are probably off.

The four measurements: 20s; 4 mm, 10s; 2 m, and 15s; 1 mm, 30s; 2 mm, would thus be conserved.

On the other hand, in the Type B row, the measurements 4s; 1 mm and 15s; 2 mm would be contested and rejected by the children, with the suggestion that they be re-done.

A different collection of measurements might produce the following line of reasoning:

Second Phase: Completion of Table; Search for Missing Values (20–25 min)

Presentation of the Situation: Instructions

In general, certain types of paper were not chosen during the communication game (the extremes being easier to communicate) At this point, students tend to notice that, and that they have no measurements on the table.

The teacher then suggests that the children complete the table by measuring the missing types of paper.

Development and Remarks

The children carry out the work in groups of five which are no longer competitive but co-operative and not restricted to the same groupings as in the previous session. (Several groups take the same type of paper and later verify the compatibility of their measurements.)

When all the children agree, the new measurements are written into the table. At the end of this phase, the table is therefore entirely correct and complete. There are many compatible measurements for each type of paper.

Third Phase: Communication Game (15 min)

The teacher suggests that the children re-play the communication game from the first activity, taking into account all the observations and corrections that they have made (large number of sheets, compression of the paper, etc.). The game is thus replayed to the great satisfaction of the children, who succeed in their task even if the teacher adds one or two new types of paper.

Results observed: These children know how to adapt the number of sheets chosen to the necessities for discrimination of their thickness (to augment the number of sheets when the thicknesses are very close). They know how to calculate pairs corresponding to the same type of paper. Everyone now knows how to use the linear model to analyze a table. A small group of the children is able to use relationships of closeness between pairs. A large number of the children have learned to judge declarations and to argue their own points of view.

Summary of the Rest of the Sequence (Session 3)

After verifying that the children know how to recognize sheets designated by their thicknesses, as for example (48 s; 9 mm), the teacher asks them to find other ways of writing each thickness, and then to rank the sheets from the thinnest to the thickest. Then she gives them (25 s; 7 mm) to rank with the others.

By the end of the session, the teacher tells the children a method of writing the thickness of a sheet: (50s; 4 mm) designates a pile of 50 sheets which is 4 mm thick; the thickness of one of these sheets is written $\dfrac{4}{50}$, read as "four fiftieths of a millimeter". To verify that they are able to use this new notation, she gives them some exercises, such as

(a) Write the thickness of papers A,B,C and D as fractions,
(b) Write the fraction designating a stack of 20 sheets whose thickness is 3 mm.
(c) Write the fraction for a sheet of paper if it takes 30 to "make" 5 mm.
(d) Can there be than one fraction designating the same thickness? Find some.

The follow up to part d is to point out that to say that 2/9 mm and 4/18 mm are the same thickness, we write 2/9 = 4/18.

Results observed: Within the context of the Situation, the children know how to find equivalent pairs. They know how to compare the thickness of sheets of paper (many of them using two methods, either specifying the number of sheets or specifying the thickness). Using these comparisons, they have a strategy for ranking pairs. They know how to designate the thickness of a sheet of paper by means of a fraction and how to find equivalent fractions.

Remark: This knowledge occurs within the Situation. It is unlikely that it is yet possible to take a question out of context and ask it independently. The results can not yet be depended upon as "acquired" knowledge, nor do the children identify them as such.

This is followed by a sequence of lessons in which the students figure out how to carry out the basic operations (addition, subtraction, multiplication and division by a whole number) in the context of the sheets of paper, then convert to the more standard notation and become accustomed to it. The learning is reinforced by some exercises involving the same kind of measuring of other objects, for instance different types of small nail, each individual nail being lighter than the available scale can measure.[5] With that well in hand, the students are introduced to decimal numbers as a type of rational number which is convenient for making approximations and has a useful notation. The basic mechanism is a Situation in which students need to take one fraction as a target and find other fractions that get closer and closer to it. They discover that this works out much the most easily if all fractions involved have denominators that are powers of ten. Once they have mastered the use of these fractions, they are given the decimal notation for them.

The next stage is to progress to fractions as linear mappings (that is, taking 5/9 of something rather than simply using 5/9 to designate a quantity). This is introduced with the Situation of the Puzzle. This Situation has been used a number of times outside of the sequence described in the manual, so that the descriptions of student responses are based on an even larger collection of observations that are those in the Stacks of Paper Situation.

Another Situation: The Puzzle

The Problem-Situation

The first Situation put to students for study of fractions as linear mappings is the following.

---

[5] For details, see Brousseau, Brousseau and Warfield (2004) pp 17–20.

Fig. 2.3 The Puzzle

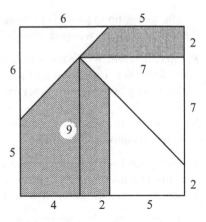

Instructions

"Here are some puzzles (Example: Fig. 2.3 above) like the ones we have been making patterns with for the past few days. You remember the nice patterns we have been making using a few pieces at a time. We want the younger class to be able to play with those puzzles, but the ones we have been using are too small for their fingers. So you are going to make some similar puzzles, larger than the models, according to the following rule: the segment that measures 4 cm on the model will measure 7 cm on your reproduction. I shall give a puzzle to each group of four or five students, but every student will do at least one piece or a group of two will do two. When you have finished, you must be able to reconstruct figures that fit together exactly the same way as the model."

Development

After a brief planning phase in each group, the students separate. The teacher puts an enlarged representation of the complete puzzle on the chalkboard.

Most of the students think that the thing to do is to add 3 cm to every dimension. Even if a few doubt this model, their conviction is not as strong as that of their classmates who do, and they seldom succeed in convincing them. The result, obviously, is that the pieces are not compatible. Discussions, diagnostics and rather heated discussions ensue. The blame first falls on the execution of the plan. Verification is attempted—some students re-do all the pieces. In the end they have to succumb to the evidence, which is not easy for them to do! The teacher intervenes only to give encouragement and to verify facts, without making particular requirements. By means of successive rearrangements, some students produce a puzzle that roughly reproduces the form of the model. The teacher, invited along with the other groups of students to confirm success, in this case suggests that the competitors use the model to form a figure with the original piece (such as Fig. 2.4) that cannot be

Fig. 2.4 A figure made with
the original pieces

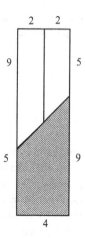

Fig. 2.5 An attempt to
reproduce the same figure
with fudged enlarged pieces

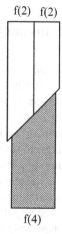

reproduced with the pieces they have made (Fig. 2.5). It is generally quite easy to find three sides, a, b, and c, such that $a+b=c$ and $f(a)+f(b) \neq f(c)$.

For a variety of social and intellectual reasons, there is a general resistance to the idea of reconsidering the initial model of the process. The result is generally a small dose of pandemonium.

When at last the children accept that there must be another law and set about searching for it, things move along more quickly, especially if one of them, or the teacher, displays the lengths on the blackboard (Fig. 2.6).

They might first find the image of 2: if 4 → 7 then 2 → 3.5. In general, at this stage, the idea is not contested, as if, as soon as the other model is rejected, this one replaces it. "We would need the image of 1." "Yes, then we could find all the others." "For that, 4 will have to be divided into four parts, and 7 will have to be divided into

Fig. 2.6 Side lengths in the
original puzzle

4 _____

2 _____

6 _____

5 _____

9 _____

7 _____

four parts as well." The model of commensuration that they have been taught would
allow them to write directly: 4 times the image of 1 measures 7; the image of 1 is
therefore 7/4. Before they get to it, though, they usually calculate using more elaborate procedures that they can construct because of their experience working with
fractions as measures. These calculations now allow success duly verified by
construction.

Summary: Analysis of the Two Situations

I shall finish this chapter with a brief analysis of the Situations it presents, so as to
point out how they fit into the general structure. I need to point out, though, one
danger to making such an analysis. Situations of Action, Communication, etc. are
by no means cut and dried. As Brousseau himself remarks, there is frequently some
overlap and some mixture of the ordering. The specific labeling could often be
argued, if the labeling were important in itself. Since labeling serves more the function of a lens than a road map, such an argument would be of dubious worth.

To start with the sheets of paper, the phase when the students are moving about,
attempting to measure single sheets of paper, trying out various sizes of stacks and
checking if the ruler can handle them is unambiguously a Situation of Action. When
they settle down to work out a system whereby the transmitting team can describe
the paper in such a way that the receiving team can recognize it, they are in a classic
Situation of Communication, where they not only must formulate what they need to
communicate, but invent for themselves a code for doing so. The game of messages
remains part of the Situation of Communication, but when the messages are assembled and the discussion of the resulting table begins, a Situation of Validation has
been reached. Their discussion covers with many iterations the issue of why two
codes are inconsistent, or a particular code is not self-consistent. After it they return
briefly to the Situation of Action, folding back in the Pirie/Kieran sense[6] to the
original activity, but now with new layers of understanding of it.

---

[6] Pirie and Kieren describe the learning process by positing a sequence of concentric circles of
increasing levels of abstraction. The student's path of learning progresses through the sequence,
but not uniformly outward. From time to time there is a return to an inner circle, which they char-
acterize as "folding back", because at each return the knowledge is in some sense thicker. See, for
instance Pirie, S.E.B., and Kieren, T.E. (1994).

The Situation of the Puzzle again starts with an unambiguous Situation of Action, and very lively action it is. Didacticians refer to this as a particularly robust Situation, meaning that even when presented out of its original context and to quite different collections of students, it can be depended on to engage the students and to maintain a high interest level that will bring them to invent the concept of multiplying by a fraction. The Situation of Action lasts as long as the students are simply cutting out pieces and pushing them together. Formulation begins when they finally accept that making repeated minor modifications of the additive model just isn't going to work for them, and begin to discuss alternative ways to proceed. Since this procedure is self-checking, this Situation does not include a Situation of Validation. That aspect of the learning of this particular concept is taken up in the Situations that follow it.

Chapter 3
A Fundamental Situation for Statistics

The Fundamental Situation

A central concept in the Theory of Situations is that of a Fundamental Situation. At first glance, this can appear to be simply a theoretical item, with major impact in the study of the field rather than in applications. Further examination reveals that in fact it has extreme and vital importance in the classroom and for students. The Fundamental Situations underlying the learning of different mathematical concepts are not simply tools of the researcher's trade—they are crucial to the entire structure and use of the Theory of Situations.

I'll start with a concise definition of a Fundamental Situation, then describe one specific but fairly typical lesson sequence illustrating the concept, then return to a short discussion using the lesson sequence to help amplify and solidify the original definition.

A Fundamental Situation is a set of conditions which provide a (possibly implicit) definition of, and an opportunity for a student to learn

- Some piece of key knowledge, with its origins and consequences, and
- The typical uses of that key knowledge.

In somewhat more depth: a Fundamental Situation not only concerns a piece of key knowledge, but is associated with a model problem presented in such a way as to permit the student to produce the solution and prove it by the exercise or creation of as complete as possible a form of this knowledge. This creation can be a direct response or the result of a "genetic" process, that is a sequence of questions being asked and answers that can be given by the student, which the situation provokes and maintains.

What follows is a description of a set of lessons constituting a Fundamental Situation for Statistics. Before beginning them, however, let us take a brief look at the intellectual context of the lessons. Probability and Statistics are subjects that provide some heavy challenges for understanding, and hence for teaching. One major challenge lies in the intersection of the two fields. In technical terms, it is the Law of Large Numbers. Less technically, it is the relationship between data, probabilities and predictions. The probability that the toss of a fair coin will result in heads is exactly 1/2. This means that if it is flipped a large number of times heads will turn up pretty close to half of the time, and if it is flipped a huge number of times, heads will turn up very close to half of the time. That much is easy to grasp. The problem is the meaning of the "large" and "huge" numbers, and along with it the issue of degree of certainty. That few people do grasp it is demonstrated by the frequency with which people state their belief that if tails have come up several times in a row it must be head's turn. Las Vegas benefits from this belief, but it is otherwise not a good one to have around.

This state of affairs supplied the general motivation for the lesson sequence described below. Motivation for specific aspects of the sequence and conclusions to be drawn from its results will form part of the discussion after the end of the description.

The lessons that follow were designed by Brousseau and carried out three times in fourth grade classrooms as a "red wire" topic—that is, one that was not part of the curriculum. They were correspondingly not taught at the standard mathematics hour, but rather in a time of the school day in which they could be carried out in sessions of highly varied length determined by content and student interest level. In all, there were 32 sessions, only ten of which lasted more than half an hour. The descriptions here arise specifically from Nadine Brousseau's perceptive observations, careful notes and excellent memory of one of the three times, An expanded version of this description appeared in Brousseau, Brousseau and Warfield (2001).

Introducing Statistics

First Session: A Curious Guessing Game

Presentation

The teacher has prepared three thoroughly opaque black cloth bags, all fairly long and narrow, and labeled them with a big A, B and C. In front of her she has a box containing a large number of black and white tokens.

Teacher: Here we have three big empty bags (*she turns them wrong side out to demonstrate*) labeled A, B and C. I am reaching into the big box and WITHOUT LOOKING I am grabbing five tokens, closing my hand well, and still without looking putting my hand into bag A. I am opening my hand. Now there are five tokens in bag A, but I don't know how many of them are black and how many white. Do you know?

Class: Noooo!

T: But you can check from the outside of the bag that there really are 5 tokens. Jean, come check…

Now I'll do the same thing for the other two bags…

Challenge

T: Now you are going to try to guess the composition of each bag. But since nobody is allowed to look in the bags, and nobody knows exactly what is in them, nobody is going to be able to tell you whether you got it right. You yourselves are going to have to convince yourselves.

C: (*perplexed—then in a murmur*) That's impossible!

T: OK then, you are allowed to look at the contents a bit, but look out: YOU MAY ONLY LOOK AT ONE TOKEN AT A TIME, AND YOU MUST IMMEDIATELY PUT IT BACK IN THE BAG!!!! That's called a draw.

First Draws

Each student comes and draws a token from bag A, shows it to the class and returns it to the bag, then does the same for bags B and C. After the first student, the teacher asks

T: Have you learned anything?

Class: (*first muttering, then more openly*): For sure there is a black token in A and in B and a white token in C.

T: Let's go on.

The class swiftly gets to the point of requesting that the tokens already drawn be listed on the board. Some of the students don't remember the results of their own draw, or the others disagree with them. (Students who learn nothing new from their draw tend to forget it right away.). The teacher makes a chart for A, B and C and records the known results. The others re-draw. Before long each line has 17 letters:

A: bbwbwwbbwwbbwwbbw
B: bwbbwbbwbbwbwwwbbb
C: wwbwwwwwbwwwwbbwb

First Declarations

T: Well?

George: In bags A and B there are more blacks than whites, and in bag C more whites than blacks.

Nathalie: No, because we might be just drawing the same token.

Jeanne: What we know for sure is that there are blacks and whites in all the bags. (*general approbation*)

Bernard: A showed 9b and 8w, B showed 11b and 6w, C showed 5b and 12w, that's sure, too.

The students' comments at this point indicate that they have established that there is a relationship between the contents of the bags and the draws, but they are not clear what the relationship is, or how to justify it. Many accept implicitly the idea that "if there are more blacks in the bag, there ought to be more blacks drawn". But as soon as it is expressed some of the others offer rational objections along the lines of Nathalie's, or "There could be a yellow one in there we are never seeing."

Others formulate objections based on the well-known spontaneous model: "If we've seen a lot of blacks, then the whites are about to have their turn."

The teacher registers some of the students' declarations by repeating them in neutral tones, but refrains from putting them under discussion. A variety of experiments have demonstrated that students have implicit models of various origins for interpreting random phenomena they encounter, and that they have no objection to discussing these opinions. However interesting the students' conversations on the subject may be, it is clear that they cannot be conclusive. It has been shown that such conversations can't even suggest working methods, and on the other hand that they uselessly revive popular spontaneous models and the epistemological obstacles attached to them. The teacher is thus well advised not to set up at this moment a free discussion which the students might regard as conclusive.

She nonetheless notes certain ones for herself on the back of the board for later. She classifies these remarks according to their object:

- About the contents of the sacks, the drawings carried out, the predictions of drawings to come
- About the values assigned to statements: true, false, certain, probable, plausible, possible, improbable,
- About methods for gaining information, augmenting conviction, arriving at a decision, and according to their form: question, suggestion, insinuation, affirmation.

She helps to extract from the students' declarations a question which closes the session:

T: How can we know whether (and convince ourselves that) when there are more blacks than whites in the bag, more blacks than whites will be drawn? Think about it—we'll talk some more tomorrow. I'll put the bags away so nobody can sneak a peek into them.

Second Session: The Numbers of Outcomes

Draws

All 17 students are present. They start over with a new series of draws.

A: wbwbwbbbwbwbbbbww
B: wwwbbwwwbwwwbbwww
C: bwwbwwwwwbwwwwbwww

A big idea: Dominique proposes that they be added up and compared with the ones from the day before.

Yesterday	Today
A: 9b, 8w	A: 10b, 7w
B: 11b, 6w	B: 12b, 5w
C: 5b, 12w	C: 4b, 13w

The students note that, as before, they have drawn more blacks from A and B and more whites from C. Yesterday's arguments are re-examined.

Student Discussion

- Nathalie: "You might always be drawing the same one!"
- Christopher: "I thought about that at home. If there were only one white in the bag it would be tough to get 13 whites in a row!"
- Claude: "When you shake it, you mix it, you can't always be drawing the same one."
- Bernard: "Yesterday there were 9 blacks in A. This time there are 10. In each of the bags there is a difference of 1."

Another Big Idea: Camille Suggests that They Make Series of 5 Draws

This proposition is instantly accepted by all of the others.

A: bbbbw
B: wwbbb

Before drawing from C, a student remarks in the relative silence

– "There are going to be more whites."

C: bbwbb

Disappointment for the student, who had had high hopes. "In C, it's not the same."

They decide to make another series of five draws, despite the discouraged opinion of certain of them: "It's not going to do any good."

A: wbwbb

– "That proves that we had it wrong, because you don't always get the same thing."

B: bwbbb
C: wwwwb

(In sum, what they have gotten is:	for A:	(4b 1w)	then (3b 2w)
	for B:	(3b 2w)	then (4b 1w)
	for C:	(4b 1w)	then(1b 4w))

The students request to be told the composition of one of the bags, and then how many blacks there are altogether and how many whites. The teacher holds out: "We must not open the bags until we know what is in them."

Third Session: Composition of the Bags and Outcomes of Draws

The students propose to continue doing series of five draws.

A: wbbbb
B: bbbbb
C: wwwww

But there are two accidents immediately noted by the students:

— "The last two (series) are false, because there are whites in B and blacks in C."

"We have to re-draw from B and C."
The rest agree:

B: bwbbb
C: wwwww

C requires another re-draw:

C: wbwww

(Noisy joy from the students, relieved by the expected event.)
But these two rejected series nonetheless contribute to the conviction of the students. Some "certainties" about the composition of C come to light. Many of them think there are four whites and a black.

The students want to do some more re-draws from C, and one of them, instantly supported by the others, proposes to recapitulate the draws of five which have come from C:

— 4w   1b
— 5w
— 5w
— 4w   1b

The session stops there.

Fourth Session: Distributions of Outcomes

Summary of Past Lessons

The teacher asks the children to describe the preceding lessons.[1]
　　"We drew from bag C to see if we got the same results."
　　"We found out what was in the bag: four whites and one black."
　　"We are going to try it on the other bags: first B and then A."

Draws from B

To reproduce their success with C the children return to doing series of five draws, this time from B. Their first results are

4b	1w
5b	
3b	2w
4b	1w
	5w

As before the students want to eliminate the series of 5b and 5w. They cross them out on the board.
　　"There are 4b, 1w or 3b, 2w."
　　"Let's draw some more—5 series of 5 draws."
So they continue with new draws:

3b	2w
4b	1w
3b	2w
2b	3w
4b	1w
2b	3w
4b	1w
3b	2w

At this point a student requests a count of all the times the draw has given a "possible composition".

4b　1w—5 times
2b　3w—2 times
3b　2w—4 times.

---

[1] The titles of the sessions are not lesson titles. They are not intended for the children. They are formulated with the aid of a vocabulary and concepts which are not those of the children, but which describe the activity in the appropriate mathematical and didactical terms.

"5b is false, we don't count it."

"Even 5w is false."

The students continue drawing from B: the next draw gives 3b 2w.

"So now we have 3b 2w five times."

An Astonishing Proposal: A Criterion

Vivien suggests: "Two of them are tied. We have to keep drawing until there is a difference of two."

The students only count the draws between which they are hesitating: 3b 2w and 4b 1w, and they ignore the others. They draw, and a bit later there are 27 plausible draws, including

3b   2w →— 6 times

4b   1w → 6 times

They decide to continue, and get 3b 2w for the next two.

Class: "That's it!!!!!! B has 3 blacks and 2 whites. C has 4 whites and 1 black!"

They arrived at this conclusion by combining the results of the two series of five draws: 3b 2w came out eight times and 4b 1w six times.

   End of the session.

Fifth Session: Frequencies

Use of the Test

The students are now working on the contents of bag A by counting series of five.

   15 series of five draws give

1w   4b—3 times

3w   2b—7 times

2w   3b—5 times

   So some of the students propose 3w 2b for the composition, using Vivien's criterion.

   Others protest:

"No! Because if you look at the results of the draws in other sessions, we always got more blacks than whites."

A student gives an argument for this opinion, implicitly combining the day's draws with those of previous days.

The class finally decides that Vivien's method is not such a sure one, which diminishes their conviction that sack A has 2b and 3w.

Frequencies

A student takes advantage of the general indecision to slip in a proposal:

Claude: "How about if three students each draw five times. We'll count up how many whites there are altogether and how many blacks and then we'll divide by three to find out how what's in the bag."

The proposal is accepted. The total draw is 10b and 5w.

b	$10/3 = 3.33$
w	$5/3 = 1.66$

The class is a little perplexed, but accepts this. One student makes a random comment that the sum is "4.99, which is almost 5."

The session ends there.

Sixth Session: Models to Know Whether the Statistics Correspond to the Contents

Review and New Start

The teacher gets the students to recall the new method they proposed in the fifth lesson: three draws of five, add up and divide by three.

Antoine proposes that they make up a new bag of which they know the contents, so they can see whether the draws resemble the contents.

Bottle Z

The teacher goes along with the student's proposition and presents new materials for making rapid draws: bottles in place of sacks and blue and yellow marbles in place of the black and white tokens. The bottles are set up so that there is a transparent cap in place of the original cork or lid, so that when the bottle is turned upside down a single marble shows its colors. She shows them how the bottles work and tells them they may put the marbles in as Antoine proposed. (The students accept the idea strongly suggested by the teacher that "bottle or sack comes to the same thing", but there are both transparent and opaque bottles and they choose the transparent ones.)

The students ask the teacher to put four blue marbles and one yellow one into the bottle. She calls it Bottle Z. It provides the following results:

5b
5b
3b, 2y
Total 13b, 2y, thus 4.33, 0.66.

The teacher then proposes that they re-do the draws from a bag to compare the results. The students decide that they should try bag C *(because they believe it contains one black and four whites.)* They draw and get:

C: 11w 4b,   which gives the frequencies 3.66, 1.33

Someone notes that the two results are equidistant from (4, 1). The class is not surprised, because they were "sure" of the composition of bag C, and were assuming the bottle did represent bag C.

Re-launching: Increasing the Number of Draws and Dividing

T: "Why do you always do three series of five?"

The students decide to five series of five with Bottle Z.

T: "Will we need to divide by three?"

"No, by five" (the numbers are not quite ideal for distinguishing their function, since they now have sets of five draws as well as five tokens.)

The Behavior of Model Z

The results are the following:

3b   2y
5b
4b   1y
4b   1y
3b   2y
Total 19b 6y, which gives the frequencies 3.8 and 1.2

They then draw twice more from Z to accumulate with the three initial draws from bottle Z. Again they find 19b 6y, hence 3.8 and 1.2. Surprise! The students don't know quite what to think of this unexpected equality, since they had begun to think that statistics changed every time.

They decide nonetheless that they need to continue making draws, perhaps to see what becomes of these slightly mysterious numbers calculated by division.

Here the session ends.

Seventh Session: Variations on the Length of the Series

Reprise

The class recalls what it did in the previous session: series of draws, first three times, then five times (each one a series of five draws).

At the suggestion of the class, they re-do the experiment with a bottle containing four white marbles and one black. This composition still represents sack C, just the way bottle Z did. Only the color of the marbles has changed. They propose to do five series of five draws.

Results:

wwwwb
wwwbb
...
wwwbw
Total   18w   7b

18w	18/5 = 3.6
7b	7/5 = 1.4

More Draws from the New Bottle

Results:

wwwbw
wwbbb
wwwww
wwwww
wwbwb

The students comment on 5w: " they're false!", but they do not withdraw them from the count and do not propose to re-do them.

A student suggests adding up the results. A discussion follows in the course of which the other students explain to him that you'd have to divide by 10.

Results:

b : 13	13/10 = 1.3
w : 37	37/10 = 3.7

Change in the Number of Series

Some of the students suggest making six series of five draws. The students continue to write up the results of each series, then count up the outcomes.

Results:

b : 6	6/6 = 1
w : 24	24/6 = 4

Commentary of the students who hadn't expected to come up with a whole number: "We hit it lucky!"

The teacher suggests in response that they operate on the 16 series of five draws carried out that day: 19b and 61w, so blacks 19/16 = 1.18, whites 61/16 = 3.81.

Postponement of the Method

The teacher suggests to the students that they also use the results obtained in the previous sessions.

But now they want to test the method on bag A and then on bag B.

A:	w	19	19/6 = 3.16
	b	11	11/6 = 1.83
B:	w	11	11/6 = 1.83
	b	19	19/6 = 3.16

The students collect all the results from the same day, but the teacher's proposal to combine the day's information with that of previous days has no impact.

The session stops there.

Eighth Session: All Possible Models…

(a) Change in materials: postponement of models and method

The ease of manipulating the bottles leads the students and the teacher to want to represent the bags by bottles. Obviously the bottles need to be opaque. For convenience the students ask that the teacher prepare, out of their sight, bottles with the same composition as the bags A, B and C, which they will then call bottle A, bottle B, and bottle C. The class has been sure of the composition of bags B and C since the fourth session. So they take an interest once again in the composition of bag A, now become bottle A. They are having trouble choosing (see session 5) between 2w 3b and 3w 2b.

A student proposes that they redo series of five draws and count how many times those two compositions appear.

They can't reach any conclusion after 15 draws, so they ask to do 150.

Result

The first composition clearly carries the day.

The class concludes: "There are 2 whites and 3 blacks in bag A, just the way there are in bag B."

And Beyond

At this stage, the teacher takes a more active role than usual. She reminds them that (in the sixth session) they had proposed another method of working: rather than counting sequences of sets of five draws, they were thinking about looking at the accumulated results of a long sequence of draws and using division to inspect that ratio. She provides directions and materials for them to carry out a large number of long sequences and compute the corresponding ratios, which they obediently do, but with decreasing enthusiasm and increasing lethargy. Eventually one of the girls suggests that they can by-pass all this by setting up transparent bottles with each of the interesting distributions of marbles and checking them instead. They do that with renewed gusto. The teacher then reminds them of the underlying motivation for all this:

> "If you are really convinced that you are right, there is no need to open the bottles, but you can prove that you are right by showing that you can predict some things about what will happen in the next draws. You could, for instance, predict that in 100 draws there will be more whites than blacks."

With this, the students' enthusiasm returns full force. They even decide independently to expand their experimental arrangements. Originally they had arranged bottles with the three combinations they had hypothesized for their three bottles: four whites and one black, three whites and two blacks, one white and four blacks. As they launch the next experiment, they decide to include two whites and three blacks as well. They also cleverly start doing their draws in sets of ten, so as to facilitate carrying out the division. In this way all groups fairly swiftly achieve 180 draws. The teacher has them not only record the numbers, but keep a running record of the ratios of blacks drawn to total marbles drawn and graph that, thus launching a visual approach to convergence.

By now the importance of doing many draws is well established, as is also the fact that the process is really cumbersome. Technology comes to the rescue in the form of a computer set up to carry out and report random draws. Thanks to its set-up, the students don't question that it represents the same process.[2] Below is the sequence of questions posed by the computer, with example answers and outcome.

---

[2] Brousseau points out (personal communication) that the children in some sense "ought" to worry, since the validity of a model should always be verified. On the other hand, in this instance jolting their naïve assumptions would only have derailed the process.

WE ARE GOING TO DRAW WITH REPLACEMENT FROM A BAG CONTAINING BLACK
   AND WHITE MARBLES. THE BAG MUST BE FILLED BEFORE WE PLAY. HOW
   MANY WHITE MARBLES SHALL WE PUT IN?
?7
HOW MANY BLACK MARBLES?
? 3
HOW MANY DRAWS SHALL WE MAKE?
?100
IN ORDER, THE COLORS DRAWN WERE
BBWWWBBWWBBBWWWWBWW
BWWWBWWWBWWWWBBBBWBW
WWWWWWWBWBBWBWWBBWWW
WWWWWWBBWBWBBWBBWBBW
   THE 100 MARBLES DRAWN WERE:
   60 WHITE MARBLES AND 40 BLACK MARBLES
TO MAKE ANOTHER DRAW IN THIS BAG, TYPE 1
TO CHANGE THE CONTENTS, TYPE 2
TO END THE GAME, TYPE 3

With this, the students come up with the idea that the larger the frequency, the
likelier the ratio of (say) black marbles drawn to total marbles is to be within a small
interval of the actual ratio of black marbles to total marbles. By way of confirmation
and application of this theory, they play a game, which the teacher sets up as
follows:

"We have a bag with 4 marbles. You are to

- guess the composition of the bag, and
- guess the value that the "accumulated frequencies" (after 3000 draws, for instance) will
  approach.

To do that, you may request draws from the computer, but you must buy them with
tokens: one token buys 5 draws. At the start you will have 25 tokens. You may buy draws
until you have a pretty good idea (a hypothesis) about the composition of the bag. At that
point you will stop drawing and stake the rest of your tokens on that composition. The
computer will then tell you the contents of the bag. If it is consistent with your prediction
(your hypothesis) you have won, and you double your stake, if not you lose your stake. You
can only bet once in the course of a round.

Be careful, if you use up all your tokens you will have nothing to stake and you won't
even be able to find out the contents of the bag (the computer requires a stake of at least one
token.) The group that ends up with the most tokens will win.

Suppose for example that you have bought 50 draws (with 10 tokens), and you think that
the bag has three whites and one black. You have fifteen tokens left which you can stake on
this composition. If you are right, you will win thirty tokens (double the fifteen). If you are
wrong you will have no more tokens. If you are not quite convinced that the bag has three
whites and one black, you may buy another 25 draws with five tokens, and then you will
have ten to stake."

Beyond that phase, the rest of the lessons are mainly concerned with formalizing
and making explicit the ideas that the class has put together.

How then does this sequence represent a Fundamental Situation? It might perhaps be worthwhile to start by observing why it is a Situation at all. For that it is helpful to look at one of Brousseau's defining descriptions. In effect it paraphrases the definition given in chapter one, but this format is more useful in our current context.

A Situation describes the relevant conditions in which a subject uses and learns a piece of mathematical knowledge. At the basic level, these conditions deal with three components: a topic to be taught, a problem in the classical sense and a variety of characteristics of the material and didactical environment of the action.

Here the topic is statistics, the problem is determining the content of the bags, and the didactical environment includes the materials (tokens, bags, bottles, graph paper, computer) and the teacher's presentation and handling of those materials.

So the question that remains is what makes this Situation a Fundamental one. Let us return first to the definition given at the beginning of the chapter:

A Fundamental Situation is a set of conditions which provide a (possibly implicit) definition of, and an opportunity for a student to learn

- Some piece of key knowledge, with its origins and consequences, and
- The typical uses of that key knowledge.

In somewhat more depth: a Fundamental Situation not only concerns a piece of key knowledge, but is associated with a model problem presented in such a way as to permit the student to produce the solution and validate it by the exercise or creation of as complete as possible a form of this knowledge. This creation can be a direct response or the result of a "genetic" process, that is, a sequence of questions being asked and answers that can be given by the student, which the situation provokes and maintains. We have already encountered one Fundamental Situation: the Situation of the Pots of Paint is a Fundamental Situation for the concept of counting.

What key knowledge is central to the Situation sequence just described? It would be easy to name various topics from Statistics—sampling, for instance, or hypothesis testing. There is a deeper idea at work, though, on which they are all based—the idea of the distinction between deterministic and probabilistic thinking. Prior to this, any situation with which the students have engaged was deterministic, that is, it involved reasoning with certainties: three apples and another five apples are unambiguously eight apples; other computations, even if more elaborate, have only one (correct) answer, and it can be known. In this Situation they are thrust into dealing with uncertainty, and thanks to the teacher's firm refusal to let them sneak a peek at the contents of the sacks, they build for themselves an understanding of the relationship between the number of observations and the state of being fairly, or very, or really extremely convinced of the validity of a conclusion. The game at the end can be regarded as a Fundamental Situation of Action for hypothesis testing, but that is a very specific sub-section of the general Fundamental Situation. If the key knowledge at hand is probabilistic reasoning, then hypothesis testing certainly constitutes one of its typical uses.

With this much analysis accomplished, a slightly more elaborate definition of a Fundamental Situation might now be appropriate:

A Fundamental Situation is a Situation which

- From the point of view of the knowledge

  - Is mathematically genetic. That is to say, it can be logically generated using some existing knowledge and it serves to generate logically other knowledge.
  - Is generic. That is, it presents a model problem to which a large number of problems included in the learning objectives can refer.

- From the point of view of learning has the potential of provoking spontaneous learning in a student or in a class using their own repertoire and without recourse to supplementary information.
- From the didactical point of view is something that can be posed to the students and resolved by them in such a way that their solution can be received by the teacher as having been constructed using only the existing institutional knowledge of the class.

It is a sad fact that after this experiment had been carried out in four different classrooms it had to be dropped, because the class period in which it was taught had to be used for other things. There is furthermore no follow-up record on the children who took part in it. They were long dispersed before they got to the point in the school system where Statistics is part of the curriculum.

Chapter 4
Gaël and the Didactical Contract

In this chapter we return to the story of Gaël and how working with him led Brousseau and his colleagues to the invention of the didactical contract. The development is described in detail in "The Case of Gaël".[1] Here I will present excerpts from the article, with occasional minor modifications, connected by my own commentary and descriptions of the intervening sections.

The most compact definition of the didactical contract is given towards the end of the article:

> We give the name *"didactical contract"* to the set of (specific) behaviors of the teacher which are expected by the student and the set of behaviors of the student which are expected by the teacher.

A more elaborate description is found in the abstract:

> It is the simulacrum of a contract, an illusion, intangible and necessarily broken, but a fiction which is necessary in order for the two protagonists, the teacher and the learner, to engage in and carry out the didactical dialectic. The didactical means to get a student to enter into such a contract is *devolution*. It is not a pedagogical device, because it depends in an essential way on the content. It consists of putting the student into a relationship with a *milieu* from which the teacher is able to exclude herself, at least partially.

The concept originated in a study carried out between 1976 and 1983 by a team consisting of Brousseau, a psychologist and a small group of collaborators and students. They worked with nine children who were succeeding in all other subjects and failing in mathematics. Gaël, aged 8 ½, was one of the children. The team initially formulated their approach to meeting his needs in terms of a mathematical Situation, but in the course of a long sequence of tutorial sessions carried out by Brousseau and analyzed by the whole team, they came to recognize a different source of Gaël's failures. Brousseau describes the major components of their work and discoveries as follows:

> 1. The Situation proposed for Gaël was directed towards replacing "constructive" (in the mathematical sense) definitions of subtraction, in which the student reproduces an algorithm

---

[1] Brousseau and Warfield (1999).

which is shown to him and which gives the desired result, by an "algebraic" definition, in which a number must be found to satisfy some condition. A difference is what must be added to some number to find some other number: $39 + \Diamond = 52$. It is the prototype of the Situations with which we explored the possibilities of replacing arithmetic by algebra as early as possible in primary school.

To clarify this aim: the arithmetic approach to which Brousseau refers is the one often referred to as "taking away". For instance, 5–2 is introduced by presenting the child with five objects and then removing two of them. Pursuing this line of questioning exclusively leads to what Brousseau is calls the "mathematically constructive" definition, with some set of place value algorithms for determining how to construct a difference like 52–39. The "algebraic" definition stems from the question "I started with five apples and now I have three. How many did I lose?"

2. In proposing to replace the construction of a term by the understanding of a relation and the search for an object satisfying it, the Situation brought sharply into evidence the paradoxical conditions of all didactical Situations, which make it simultaneously necessary and impossible to maintain an effective **didactical contract**. This concept had its birth in the course of the experiment.

3. Finally, this experiment brought to light the relationships and irreducible differences between the didactical, the psycho-cognitive and the psycho-affective approach to a teaching situation.

We begin to get an impression of Gaël's mathematical functioning (or lack thereof) in the first session. The tutor (Brousseau) brings up a question which Gaël had answered wrong in class: **In a parking lot there are 57 cars. 24 of the cars are red. Find the number of cars in the parking lot which are not red.**

First Gaël says "I am going to do what I learned from the teacher" and adds 57 and 24. The tutor encourages him to draw the cars—but not all of them, because it would take too long. So Gaël draws a rectangle and puts the number 57 in it.

T (the tutor):   "Is that all of the cars?"
G: "It's all of the cars that aren't red."
T: "Only the cars that aren't red?"
G: "It's all of the cars and they aren't red."

At this, the tutor abandons various more sophisticated plans and has Gaël make 57 tally marks in the rectangle in sets of 20. At the end, Gaël still thinks he needs to draw a 24 more for the red cars. This gives rise to the first question for the tutor:

Here one can see that he has difficulty in envisaging that there is **only one set** *of cars, with two properties: being in the parking lot and being red. For him, the second property necessitates a second set, and even though he admits that the second set also has the first property, he can't yet conceive of its being a part of the original set. Is it because he hasn't analyzed the statement of the problem, or because he can't use the operation of inclusion?*

Eventually (by now totally guided by the tutor, with specific instructions for each step) Gaël succeeds in designating the red cars and verifying that the resulting diagram corresponds to the problem statement. Gaël then counts the non-red cars and gets 31. A re-count yields 33, but the same conversation makes it clear that checking by using an operation rather than counting is not in Gaël's sphere of ideas.

Eventually, the tutor sets up the corresponding addition problem (24 + __ = 57), with a lead-in such as to make the problem one that Gaël can now handle easily.

At this point there is clearly a morass of question marks about where Gaël's actual comprehension lies. The tutor first does a little checking with addition and subtraction problems using small numbers and discovers that the issue of inclusion continues to give Gaël difficulties. He then tries out some very basic tests used by Piaget and Gréco to make sure that Gaël has the necessary mental structures for dealing with these problems.[2] Gaël passes with flying colors, but an extremely revealing exchange takes part at the end. After having given with complete confidence the correct answer to a question involving the length of a stick, Gaël proceeds to back down instantly when told "a kid told me a few minutes ago that this one was shorter."

This ends the session, which Brousseau analyzed as follows:

Among the questions arising from Gaël's behavior there was one which had to be checked out swiftly. All mathematical activity is supported by operational schemes of the subject which, according to Piaget, are not learned in the strict sense, but constructed in the course of development. In the case of the red and non-red cars, we could see that it was absolutely necessary to assure ourselves that Gaël really did have the operational structure of inclusion.

We carried out this testing, and the results assured us at least that Gaël's repeated failures to understand the problem could not be explained by gaps at the level of his logico-mathematical structures. Manifestly, he had the operational schemes necessary to solve the problem proposed. How then to explain his behavior in the course of the session?

It was clear the answer would not be a simple one. Looking for it required that we take as an object the **relationship** between a single subject, whose current relationship with the world was the result of a lot of past history, and a didactical situation which itself was quite complex.

In this perspective, one of the tests gave us something to start with. The immediate cause for surprise was Gaël's complete incapacity to maintain his conviction in the face of anyone else's contradiction. One counter-proposition was enough to produce doubt right at the moment when to all appearances he was feeling completely convinced. A characteristic of the child thus appeared which we had already encountered in the course of a psychological examination on a more general front: flight from any possible confrontation and the avoidance of conflict at any cost by taking refuge in a position of dependence and submission.

It seemed to us that this might well have an impact on Gaël's relationship with mathematical knowledge. In the area of knowledge there is, in effect, an attitude in which dependence offers the non-negligible benefit of a kind of security: knowledge is always somebody else's knowledge which one has only to appropriate. Thus, one eliminates the risk of being the one put into question in a debate about truth. There is no need to offer any reason for what one takes for truth other than the invocation of the authority to whom one refers. (Gaël says "what I was taught", "what the teacher says I have to do".)

But the price of this attitude is an incapacity to conceive of a process of construction where knowledge might be the result of a confrontation with reality, and in which the subject becomes the author of his own knowledge. Mathematical knowledge thus risks being simply a ritualized activity of reproducing models.

---

[2] To test for his ability to handle the concept of inclusivity, they gave him eight beads, five of them red and three green, and asked him to say whether there were more beads or red beads and justify his answer. To test for the concept of commutativity they showed him three rods of lengths A, B and C, laid out so as to make it clear that $A + B = C$. They then removed A, slid B down to line up with the left hand end of C and asked him where the end of A would be if they put it down beside B. He instantly replied that it would still come to the end of C.

With this information and analysis in hand, the group began to contemplate how to approach the second session. They looked first at the possible results of proceeding in a relatively standard way:

A classical approach in dealing with children in difficulty consists of identifying the errors or mistakes that they make, and if they are repeated, interpreting them as anomalies in the development of the child, or gaps in their acquisitions of knowledge which need to be remedied because "they are going to make the child unable to progress in mathematics."

For example, we observed that Gaël frequently wrote S for 5, or wrote 12 for 21, which could be interpreted as a lack of **spatial structure** or even trouble with **spatio-temporal** perception. In the same way, Gaël's difficulties in connecting his drawing with the text of the problem could be classified as malfunctionings of the **symbolic function**.

This classical analysis leads to a search for remedies in the form of exercises "of the same type" in the sense of these gaps: exercises in spatial structure, etc. The approach we are trying here is very different: it's a matter of working at the level of the learning Situation and manipulating its characteristics so as to obtain the desired changes in attitude. For that we will use the Theory of Situations. This theory studies, as its principal object, the conditions of the *milieu* which make the behavior of the subjects and the manifestation of knowledge necessary and plausible.

Gaël's relationship with knowledge—at least the knowledge involved in the classroom—is strictly superficial. His habit of avoiding problems and keeping his distance lead to stereotypical actions of a purely "didactical" nature—that is, centered entirely on the relationship with the teacher without mobilizing any assimilation schemes even though he actually does have them at his disposition. Gaël accommodates himself to a set of institutionalized relations which on his side call forth only rituals that do not engage him at all. It seems possible, thus, that all of Gaël's attitude during this first session is the consequence of an accord between the habitual didactical situation in the class as he perceives it and his defensive relationship with knowledge.

One cannot maintain the position that the didactical situation that Gaël habitually encounters is the sole cause of his failures in mathematics. If it were, why would other children, no better armed than he on the cognitive front, succeed? All we can think is that he finds this situation a convenient one in that it lets him escape the effort of constructing knowledge. And he can escape it all the better for his manner of dealing with adults—his own particular social attitude made up of sweetness and submission, which defuses all criticism, leaving him forearmed against any form of conflict with the teacher.

Why does this cause failure? Because, if the habitual didactical situations permit learning in the traditional closed conditions, it is because engagement with knowledge is replaced by another type of engagement—one which features the learning process itself. Learning badly, not knowing, making thus and such an error are all forms of running afoul of the will of the teacher, of being in conflict with him. From there out, the student can only escape conflict and all the resulting difficulties by building something which will play the role of knowledge and of learning.

Now Gaël, we would say, escapes this debate by the extent to which he disarms all conflict by a total absence of aggressiveness. In student-centered engagement, the conflict effectively feeds on itself: the teacher's aggressiveness calls for a habitual response of aggressiveness from the student, which in turn feeds the aggressiveness of the teacher, etc., with the student having no way out except by producing the expected results. Gaël, for his part, doesn't join this game. His profoundly submissive attitude disarms all hostility ("He's always ready to acknowledge his mistakes and he's really sorry" says the teacher.)

Based on this analysis, they decided that

the essential issue will be to introduce a rupture in Gaël's conceptions of a didactical situation by offering him a Situation which will require of him anticipation, prediction and the undertaking of responsibility.

To this end, the researchers prepared a bag containing 52 geometric objects: small and large circles and small and large triangles. Repeatedly, all the objects of one shape (say, the small triangles) would be pulled out and counted. Gaël's job was to give the number remaining in the bag. But he wasn't merely to give the number—he was to bet on it. In fact, he had nothing to lose on the bet, and the caramels that he won were purely hypothetical, but nonetheless the bet had a major impact on his reactions. He would check his first answer (generally pretty much of a guess) and use the results to attempt to replace it:

> The number of givens that Gaël paid attention to suddenly increased a lot, and he could only handle the cycle once through: choice, anticipatory verification, rejection, new choice. The bet was a period when the tension relaxed—a pleasant moment of pretending to think, hesitating a little, then deciding and solemnly shaking hands with the tutor. Then with a bit of slightly feverish excitement the sack was opened, the count was reckoned, the resulting number was compared with the prediction, while the tutor looked on dubiously with furrowed brow, simultaneously sorry, encouraging and comically powerless. The bets had to remain reasonably frequent to maintain the child's pleasure—they were the real source of gratification.

Variations on this one game occupied the entire second session. The one additional feature that the tutor was able to smuggle in, in a way that escaped any hint of being something Gaël would feel he was "supposed" to do, was that a proposed solution could be verified by counting upwards. That is, if seven small triangles were visible, and Gaël suggested that this meant there were 43 objects remaining in the bag, he could count off 44, 45, 46, …, 50, one triangle at a time. As he mastered this tactic, Gaël was able to carry through increasingly long sequences of predictions and corrections. This led to the following comments:

- For a start, it seems perfectly clear that Gaël is completely capable of entering into a Situation of Action. He accepted the rules of the game, which consisted of taking charge of an objective and of the means of verifying by himself that it had been achieved, of hazarding solutions and of checking them against the state of the *milieu*. He took over the search for a good answer, rejecting contradictions and inadequate solutions himself. He took pleasure in the game of predicting and verifying even when he didn't win.
- He voluntarily entered into a state of anticipation. This last point is very important for more than one reason:
  - Anticipation is the first step towards creation of a theory and passage to an experimental basis: the subject gives up the procedural mode, with only direct interactions with the *milieu*, and the trial and error method and considers his actions in a wider context.
  - Anticipation requires the existence of at least an implicit model, true or false, on which it is based and the expression of which permits it to be put to the test. Here the model is the relationship: number known + number tried = 52, and it seems dependable enough to Gaël to permit him a rapid simulation of the experiment.

It is interesting to note that Gaël did indeed master the steps of anticipation: we wanted him to interest himself in the material nature of the givens of the problem statement; we produced the collections in question with a bizarre and captivating ceremony so as to augment the affective, perceptive and sensorial weight of the search for a solution. Now, investment in anticipation is in a way antagonistic to investment in action, in the sense that it assumes at least a provisional refusal. When Gaël takes on the responsibility of verification, he renounces the pleasures of action, of decision, of betting, of the game, to replace them by calculations and simulations. But in any case, it should be remarked that anticipation inherits to a certain extent the motivation associated with the situation that it simulates. Gaël experiments with his predictions with the little shiver of pleasure recalling the one he feels at the moment of betting.

Finally, the successive consideration of several possible predictions in the course of a single bet, and the fact of writing them permits the examination **at the same time** of diverse choices and consequently the choice of a strategy based on the structure of this universe of possibilities. This is an essential step in changing the process of predicting from a yes/no guess to a logical selection from a amongst collection of possibilities.

Gaël is thus capable of entering into all these phases of the dialectic of action, of producing and checking implicit models. He probably accomplished this activity naturally. His stereotypical attitudes which could be observed in class and his tendency to search for easy answers by interpreting the suggestions of the adult are thus an effect of his manner of dealing with the didactical situation.

We were able to see here that this situation could be altered. The choice of an appropriate Situation did indeed produce the rupture we envisaged, with the efforts predicted. To be sure, this "incidental" rupture had not yet changed Gaël, nor his relationship with knowledge. It was partly obtained by making use of his major fault: the desire to seduce the adult and maintain affective and playful relations with him. The challenge for us now was to establish more firmly in Gaël's scheme of things the possibility of interacting with knowledge itself. It couldn't be accomplished by having his teacher change her ways, because the initial information given to the researchers indicated that most of the children in her class were responding fine.

The third session went on in the familiar format, although Gaël succeeded in startling the tutor early on:

The tutor takes up the guessing game from the last session without any modifications: there are 56 pieces which Gaël counts and puts in a bag after having written the number on a sheet of paper. The tutor then takes 10 big circles out of the bag, has Gaël count them, then puts them, together with the first bag, in a larger bag. The question is how many pieces there are in the inside little bag.

Gaël thinks a bit, counts to ten and says "5!" The tutor shows him the 10 circles, then shakes the other bag and asks if he really believes there are only five in there. Gaël smiles, blinks and shakes his head, acknowledging that he has made a mistake. In fact, Gaël has just reproduced his habitual mode of response: he counts to ten—stereotypical behavior—then since he has to give an answer, gives a random one.

Gaël then follows up in a mutter: "56…then there are 10…(he counts to 40), I'm up to 40 and 10 are taken away,…that makes 40!" In fact, he hasn't changed procedures, he has just given a more plausible number to live up to the adult's expectations. The fact that the number to subtract is ten, the magic number, may contribute to this disarray.

The tutor then reminds him of the principle of verification of the statements during the last session: "We made a bet and we checked if we were right"…without, of course, saying what operation was used. While he is doing so, the child, smiling, thinks, then exclaims: "46!"

It seems clear that he is sure of his answer. It seems possible that reminding him of the conditions of the situation has been enough to make it possible for him to check his answer and thus to produce a correct one. It also seems possible that Gaël was playing a subtler game, throwing out provisional answers to gain time to think it over, or even simulating his habitual response to tease the instructor. In any case, this scene demonstrates Gaël's need to "decorate the silence".

T: (slightly jolted by the rapidity of the response) "How come you're saying that makes 46?"

G: "Because I know that you take 5 from 10 and that makes 4 and that leaves 6 so that makes 46."

T: "!!!"

Doubtless the way to translate this answer is: from 5 tens we subtract one ten (formula backwards), and that leaves 4. In 56 there are 5 tens and 6 ones; these 6 ones together with the 4 tens give us 46.

After explaining his answer (46), Gaël sets out to verify it. But first the tutor asks if it isn't possible to know how many pieces there are without opening the bag. Gaël's reply: "Oh, no. There is no way of knowing."

The tutor then reminds him of the method used the time before, when they considered that the bag had a certain number of pieces and then added onto this number the number of the pieces spread on the table: if no error had been made, the resulting number was the total number of pieces. Gaël uses this method and figures that in fact, this time he has not made a mistake. To be absolutely sure of the result, they are going to empty the bag and count all of the pieces, after betting. Gaël says he is not "absolutely sure" of winning.

Each one counts some of the pieces, with the tutor making stacks of 10 pieces. At the point when between them they have counted 40 pieces, Gaël stops and says "Oh, I know I have lost." but the tutor encourages him to continue, and at the end of the count he realizes that he had it right.

The session continued with the betting game as its basic structure, and incorporated also some mathematical adjustments designed to solidify Gaël's understanding of subtraction. After it, the study team came to the following conclusions:

The return of the "game" of the bags with slightly simpler givens permitted Gaël to rediscover the initial schema and produce the expected reasoning. We also discovered difficulties and errors of his that we had already known about. Repetition of Situations of this sort would unquestionably permit us to lead him to correct his writing errors, to know numeration well and to give some meaning to subtraction problems—especially since the tutor had managed to develop a pleasant relationship with Gaël. There is no doubt whatever that to please an adult, Gaël would identify what is pleasing to the adult, manifest the expected behavior and simulate the desired acquisitions. This would give him the time to cement the affective relations which do not deal directly with mathematical knowledge and break off the ones that might make him uncomfortable. But that is precisely the problem: the price of his progress would be the reinforcement of exactly what led to his failure in the first place.

From the beginning of the sessions we were struck by a profound tendency of Gaël's to give more or less plausible spontaneous answers. He seemed incapable of suspending his highly impulsive reactions for a time in order to reflect, assemble information, and slowly construct the necessary inferences. In effect, he was willing to go to considerable lengths in order to avoid dealing with certainty. This tendency was so strong that, sure of the result, he would still try, by a genuine denial, to abolish the character of certainty from his reasoning: "Oh, I know I have lost!" The question was what this meant and what to do about it.

What we could assume was that given that he was quite capable of carrying out the relevant operations, this attitude had some benefit to him, either in providing some satisfaction or avoiding some unpleasantness. We were therefore facing one of the most delicate points of a didactical intervention of this type. The mathematical failure appeared here as a symptom of something else, that is, it had its origins in the whole structure of the student's mental organization and his internal equilibrium.

From that point, if the intervention had aimed purely at remediation geared to the disappearance of the symptom by putting in operation strategies focused directly on the symptom itself—helping the child to use reasoning, etc.—we would have set him up for failure.

Given that his attitudes played a role of defense against anxiety, the counter-investment of the child would have come into play and the tutor would have been powerless to make a profound difference in Gaël's avoidance conducts.

The question, then, was how to deal with this delicate situation. One possible approach would have been the purely psychological, but it had very little to recommend it in the current circumstances. The decision then was

to restrict ourselves to the domain of the learning of mathematics, but to consider the child's mental functioning in its relations with his psychological organization and equilibrium. It was not a matter of teaching the child to reason, but of giving him, in the context of the mathematical activities planned, the occasion and the desire to do so.

This led to some deep digging into the possibilities:

What didactical means permit reasoning to function? For the moment the question was not their content or their methods, but their motivation and especially, given Gaël's practice of avoiding certainty, the role played in them by conviction.

Technically, the student's level of conviction manifests, confirms and strengthens itself in each of the four types of didactical Situation, in a different form specific to each. In Situations of Action, conviction is affirmed by the confidence of the student in his anticipations. In Situations of Formulation, the key ingredient is the distance the student must put between himself and the idea in question in order to make it an object of discussion. The resulting distance between what is said and what is thought, between a proposition and its implicit truth value, between what is explicitly predicted and what happens, raises the issue of the conviction of the speaker. In Situations of Validation, judgments and proofs are expressed and tested. Part of the convention of this Situation is that the subjects exchange opinions about a fact and engage themselves in the defense of those opinions. In general, the Situations of this nature set up a proponent and an opponent, both of whom are students, in such a way that they elaborate a system of proof founded on internal conviction and not on authority. The search for truth is a demanding activity which the searcher needs to maintain strongly enough to refuse to be convinced by anything other than personal judgment while nonetheless never refusing to examine any other argument. The searcher must resist authority, seduction, rhetoric, intimidation, social convention, etc. And when it becomes clear that his opinion is false, he must also be able to change his mind, retract and again resist the same difficulties. These difficulties are legitimate, and tend to the establishment of reasonably stable truths which become part of the internal structure of the persons who profess them. Practicing Situations of Proof or Validation permits the subject to construct an interior interlocutor with whom he can simulate debates that clarify his conviction.

The conviction takes on a different form in Situations of Institutionalization. Here the teacher is taking the ideas that have been formulated and argued about and agreed on by the whole class, formalizing and solidifying them, and possibly connecting them to the ideas formulated by society at large. For students who have been thoroughly engaged in each of the previous stages, it adds an extra level to existing well-founded conviction. For a student like Gaël, however, it would provide a blind conviction on the basis of authority—not what he needed.

In the "problem-solving situations" in class in which we had been told that Gaël participated well and had some bright ideas Gaël could say whatever passed through his head because he had confidence that the teacher would extract from it what she wanted. Gaël could say things which he "saw" as true without having to affirm that they were. This attitude may have been encouraged or even provoked by the group inquiry methods often used in classrooms.

With all this to consider, the question was what Situations would be appropriate in the case of Gaël, and which ones could be offered to him.

A classic method would have consisted of "exploiting" the Situation of Action created in sessions two and three, that is, of pushing Gaël to take part, formulate declarations, affirm them, retract them, in a relationship dual to the adult's. The tutor would extract a meaning from each action of Gaël's or have him extract one. For instance, he could repeat the betting situations and insist "You have to be certain! Are you sure?" It was clear that this method could not succeed.

On the other hand, the tutor, being alone with Gaël, had no way of organizing genuine Situations of Validation in which the child was supposed to state his convictions to an equal. He might have simulated them—and this perhaps could have been valuable in that a certain amount of identifying with the tutor might help Gaël out of his "babyish" role. But there was a danger that his tendency to stay in a superficial and playful relationship with the adult could destroy any possibility of a debate about knowledge. The playful attitudes consciously used by the tutor risked being "recovered" to reproduce the fundamental dilemma pointed out above.

At this point, although Brousseau does not so describe it, he as tutor had set up with Gaël a very definite contract, one which certainly differed from the one Gaël had with his classroom teacher. Gaël had taken on a certain amount of responsibility and had learned to deal with anticipation, and they had a pleasant, slightly playful relationship. The problem was that the researchers could see that this contract was self-limiting. As it stood, the tutor was not in a position to take Gaël to a new level of responsibility, and hence of learning at a non-superficial level.

The tutor therefore needed to accomplish a new modification of the "didactical contract", reintroducing some demands. In fact, it was to be hoped that a sequence of ruptures could be introduced; in alternation, the tutor could present himself as a partner, as an accomplice in a game, or else as an examiner who expected something of Gaël and who told him what that was. It seems evident in any case that the object of the teaching needed to stay hidden to avoid the immediate acquiescence and submissive behavior we have talked about. In the case of Gaël, what position could this tutor-partner occupy? We knew how much Gaël was dependent on affective climate, to what point his attitude was determined by the other's. An attitude of excessive affective neutrality would send him right back to the stereotypical reactions of false intellectual activity; too great a connivance would permit him a playful attitude where he could behave childishly, avoiding responsibility. It was essential to find the right distance. What the tutor was aiming for was the right alloy of complicity, where the mediation of knowledge and its own demands could constantly intervene.

The mechanism for arriving at this "right distance" appeared in the middle of the next session. After starting off with a variant and elaboration of the "game" of taking things out of the bag and predicting what is left in it, in which some organizational devices were introduced and Gaël learned their use but progressively lost his sparkle, the following took place:

They take up the game again, but this time it is the tutor who takes the objects out of the bag, and furthermore he includes a new element in the rules: they are going to play "Liar!" Gaël doesn't know what that one is about, so the tutor explains:

– "I am going to take out objects, and when I finish I will say 'There are this many greens, or that many blues, etc. (in the bag)', and if I am wrong you say "Liar!" If I succeed in lying, I win, but if you succeed in trapping me when I lie, you win!"

They try a first round: the tutor takes out a small bunch and states:

– "There are 6 blues in the bag."

At his request, the child writes the number on the chart and checks by counting backwards, starting with 10 and counting off the blue objects in sight. He agrees with the number stated.

- "There are 6 reds", the tutor next states.
- "6 reds" repeats Gaël, "6,7,8,9" (he adds in the reds spread on the table): "Liar!"

Gaël says this word with a little bit of concern and a lot of pleasure. He smiles. It needs some audacity, even though he knows that he had the license to do it. Under the fiction of the game, Gaël enters the other role, that of interior interlocutor mentioned earlier. The passage from one position to the other, from declarer to judge, from liar to person responsible for the truth…and above all the possibility of passing from one role to another offers Gaël the means for a symbolic rupture with his earlier position.

The verification is done and yes! Gaël is right and the tutor has been caught.

The sessions continued from there—four more of them, in fact—but in retrospect it appeared that this last session was the decisive one. Gaël returned to his classroom with an attitude towards mathematics sufficiently changed so that he was able to integrate well into his class and rapidly fill in his mathematical gaps.

Brousseau summarizes the aims of this intervention:

a) At first to establish a climate of confidence: an agreeable dual relationship which nonetheless took account of the difficulties at issue.

b) At the second stage, to make use of this relationship to propose to Gaël some didactical situations in which knowledge was not to be found in either the discourse or the desire of the teacher, but rather in a relationship with the milieu. These interactions needed to be motivated by the desire of the child himself, and to lead him to take on decisions specific to the knowledge to be mastered: experiment, decide, search, etc.

c) At the third stage, by new ruptures with the didactical contract, the issue was to get him to give a "price" to truth, and possibly to prefer it to the comfort of a consensus: to choose, for instance, to verify a result despite the discomfort of acknowledging an error. It was obviously not a matter of producing a moralizing lecture on the subject, but of obtaining these behaviors in an effective way. We tried to get him in the habit of defining himself, recognizing himself and pleasing himself as a constructor of knowledge and the person responsible for his convictions, faced with the facts or faced with others around him. We wanted him to experience mathematical activities not as "the discovery of his mistakes", "the recognition of failure", or "the revelation of his sins" but as an exercise in equilibration, liberation and foundation of "myself".

The reader should not be led astray by these formulations. It is not a matter of psychotherapy but of *Didactique*, that is, of specific activities intentionally organized with a view to the acquisition of specific knowledge. But it is necessary to be aware of the psychological dimension of these interventions.

Brousseau then goes on to place the study of Gaël in the context of the general study of students in selective failure (i.e., failing only in mathematics), and finishes with the paragraphs in which he first introduced the concept of didactical contract:

In the course of a session whose objective is to teach a student a specific piece of knowledge (*a didactical Situation*) the student interprets the situation presented to him, the questions asked him, the information given him, and the constraints imposed on him as a function of whatever it is that the teacher reproduces, consciously or not, in a repetitive way in his teaching practice. We are particularly interested in whatever amongst these habits is specific to the knowledge being taught: we give the name *"didactical contract"* to the set of (specific) behaviors of the teacher which are expected by the student and the set of behaviors of the student which are expected by the teacher.

Present in this question, this "contract" rules the relations of teacher and student on the subject of projects, objectives, decisions, actions and didactical evaluations. It is the "contract" that specifies the reciprocal positions of the participants on the subject of the task, and that specifies the deep meaning of the action under way, of the formulation or of the explanations furnished: what do we need to know? How are we going to tell if we have succeeded? What are we supposed to do if we haven't succeeded? What should we have known in order to succeed? What are we supposed to say? What else could we have done? What would have been a mistake? What are we supposed to learn? How can we learn it? How can we remember it?, etc. It is the "contract" that explicitly fixes the role of knowledge, of learning, of memory, etc.

It is through the rule of decoding didactical activity that scholastic learning passes. One can think that at each instant the activities in a process depend on the *meaning* he gives to the situation proposed him, and that this meaning depends heavily on the result of the repeated actions of the didactical contract.

The didactical contract thus presents itself as the trace left by the habitual requirements of the teacher (requirements that may or may not be clearly perceived) about a particular situation. The articulation between the habitual or permanent and what is specific to the knowledge aimed at vary with the student and the knowledge in question; certain didactical contracts favor the specific functioning of the knowledge to be acquired, others not; certain children read (or don't) the didactical intentions of the teacher, and have (or don't) the possibility of extracting a favorable learning situation.

Thus was launched the concept of the didactical contract. More details of the sessions with Gaël and of some of the commentary can be found in "The Case of Gaël" (Brousseau and Warfield 1999).

Chapter 5
Glossary of Terms Used in *Didactique*

In the hope that the reader will be inspired by this Invitation to go and do further reading of the book and articles about *Didactique* cited in the bibliography, I am including a collection of some of the key terms. The first twelve are ones that have appeared in the course of the previous chapters, and for those I will produce a brief description and a reference. After that come an additional five that I include because they seem to me to have the potential of easing and clarifying the reading of *Didactique*.

1. *Didactique* of mathematics

A research field whose central goal is the study of how to induce a student to acquire a piece of mathematical knowledge. From its beginnings in the late 1960s, it has grown into a field occupying many hundreds of researchers in a number of countries. In particular, active research is being done in Spain, Germany, Italy, Brazil and Mexico. The number of topics within the field has also expanded. One of the major objectives is to maintain high standards of scientifically sound research as the foundation for its theories. It models itself as closely as possible on the experimental sciences, progressing from conjectures to hypotheses, to experiments designed to test the hypotheses, to careful analysis of the resulting data.

I have chosen to leave the name in French because the word "didactic" in English tends to be a term of disapproval, and when used to describe a style of teaching, describes roughly the antithesis of the kind *Didactique* generally studies. Recently there has been a move to use the term Didactics, which has the merit of being linguistically parallel to Economics and Mathematics. This could prove to be the wave of the future.

2. Theory of Situations

This is at the heart of *Didactique*. It theorizes that all of mathematics can be taught by putting students into appropriate Situations. Experimentation on such Situations has occupied many didacticians, and Brousseau in particular, for several decades.

V.M. Warfield, *Invitation to Didactique*, SpringerBriefs in Education 30, DOI 10.1007/978-1-4614-8199-7_5, © The Author(s) 2014

3. Situation

The central notion of the theory: A Situation describes the relevant conditions in which a student uses and learns a piece of mathematical knowledge. At the basic level, these conditions deal with three components: a topic to be taught, a problem in the classical sense and a variety of characteristics of the material and didactical environment of the action. A number of these have been illustrated in the preceding chapters, from the relatively small scale Pots of Paint and Race to Twenty in Chap. 1, to the multi-lesson Black and White Marbles in Chap. 3.

I have chosen to capitalize the word Situation used in this way in order to distinguish this use from the non-specific, standard uses of the word situation.

4. *Milieu*

All of the pertinent features of the student's surroundings, including the space, the teacher, the materials and the presence or absence of other students (or another student.)

5. Situation of Action

Usually an initial phase within a Situation, in which the student is acting directly on the *milieu* and acquiring a sense of it. An example is the phase of the Race to Twenty when students are in pairs, simply playing rounds of the game without further direction (but with a lot of intention on the part of the teacher and the researchers!)

6. Situation of Communication or Formulation

Another phase of a Situation, one that tends to follow the Situation of Action. It is the one that provides the student with a need to articulate the ideas that have been developing implicitly.

7. Situation of Validation

This is the phase in which the student needs to convince someone else (and often thereby herself) of the validity of the ideas she has developed during the Situation of Formulation.

It should be noted of all three of the above Situations that despite the fact that the wording can make them appear very cut and dried and sequential, they frequently overlap considerably and merge into each other.

8. Situation of Institutionalization

This one has rather a different status from the others, and does have a specific, late place in the sequence. It is the Situation in which the teacher takes the ideas the class has developed, reviews them, shapes them up and if necessary provides them with labels. When it is pertinent, she provides the bridge between the class's production and the concepts and terms accepted by the world at large and in particular the standard curriculum.

This Situation was formulated somewhat later than the others. Its history is given in Chap. 1.

9. Fundamental Situation

An underlying hypothesis for the Theory of Situations is that for any mathematical concept there is a Situation into which a student can be put that will cause the student to construct that concept, and place it correctly within her mathematical scheme of things. It has been remarked that the truth or falsehood of this hypothesis is almost beside the point—what matters is that it be accepted as true in order to maintain the research field. Such a Situation is called the Fundamental Situation for that concept. The pots of paint Situation at the beginning of the Introduction is a Fundamental Situation for counting, and the black and white marble Situation that occupies all of Chap. 3 is Fundamental to Statistics.

10. Devolution

The transfer of power and authority from one body to another. In France, it tends to refer to the action by the king of designating that some particular piece of power, all of which was his by divine right, was now being handed over to a legislative body. In these less regal situations, the issue is more one of responsibility than of power. The current default assumption in school mathematics is that the teacher is responsible for everything, including choice of knowledge to be learned, choice of the manner in which that knowledge is to be taught and seeing to it that the student actually learns it. The student is simply responsible for carrying out whatever tasks the teacher assigns. On the other hand, it is generally accepted that any learning that is done in this format will be of a very shallow nature. Hence, the central idea of a Situation is that a student will take on the responsibility of solving some problem or carrying out some task the achieving of which will require him to use his existing knowledge to develop some concept or create some knowledge for himself. Therefore a necessary ingredient of a Situation is a devolution of responsibility from the teacher to the student.

This gives rise to something resembling a paradox: if a student accepts and responds to her teacher's authority, she won't learn, and on the other hand, if she doesn't accept the Situation presented to her by her teacher she won't learn either. What must be successfully negotiated is the devolution of the responsibility for the problem at the heart of the Situation.

11. Didactical contract

An implicit set of expectations that the teacher and the student have of each other. This is not a question of when students are allowed to sharpen their pencils, but of expectations at the level of learning. Since it is implicit, it tends to become visible only at moments when it is broken. For instance, when a student fails to solve a problem that has been posed, he tends to feel that the teacher has broken her "contract" to provide problems easy enough to be done, and a teacher tends to feel that the student has broken his "contract" by not trying hard enough. Since neither "easy enough problems" nor "trying hard enough" can be defined, they certainly cannot form part of an explicit contract.

In the Theory of Situations, the contract issue becomes central because of the need for the devolution described above. In effect, the functioning of a Situation involves repeated ruptures and renegotiations of the didactical contract.

The origin and some of the impact of the didactical contract are discussed in Chap. 4.

12. Obstacles

There is a common illusion that if the teaching is good, the amount a student knows will smoothly and quietly expand, like a well-managed helium balloon. One of the reasons that this is an illusion is the existence of obstacles of various natures. The best known is the epistemological obstacle, which consists of something that is learned in a narrow context, but that becomes false in a wider one. One of the clearest examples of this concerns numbers: a child's first encounter with numbers is, and must be, the whole numbers. The notion of number therefore comes to include things like "for every number, there is a unique number directly before it and a unique number directly after it." This is absolutely true for integers and absolutely false for fractions and decimal numbers. It therefore constitutes an unavoidable obstacle in the learning of the latter.

Chapter 1 contains some further discussion of obstacles.

13. S-knowledge and c-knowledge

The first decade or so of efforts at translating *Didactique* into English produced a variety of interesting linguistic challenges, each of which had to be tackled individually. One of the stickiest was "knowledge". Our language, although generally rich in words carrying fine distinctions, fails to supply us with a pair of words to correspond to the French distinction between "*savoir*" and "*connaître*". Both are translated as "to know". Likewise the nouns associated with them, "*les savoirs*" and "*les connaissances*" are generally both translated as "knowledge". At times this is fine, at other times it is a problem. The latter is often the case in dealing with *Didactique* where the words are frequently and highly intentionally distinguished. In the following paragraph I will attempt to make the distinction clear. After that I will describe the method we now use to attempt to make the distinction.

At a first pass, the verbs can be thought of as "to be familiar with" (*connaître*) and "to know for a fact" (*savoir*). For some examples the distinction is clear and useful: "*Connaître*" a theorem means to have bumped into it sufficiently often to have an idea of its context and uses and of more or less how it is stated; "*Savoir*" a theorem means to know its statement precisely, how to apply it, and probably also its proof. On the other hand, when it comes to an entire theory, with a collection of theorems and motivations and connections, what is required is to *connaître* it. *Savoir* at that level is not an available option—but on the other hand, no real *connaissance* is possible without the *savoir* of some, in fact of many, of the theory's constituent parts.

The corresponding distinction exists between the two words for "knowledge", with the additional complication that each of the French words has both a singular and a plural form.

Before offering a solution, I propose to give an example of a way in which having the two words is both thought-provoking and a material aid in analyzing what's going on. Currently in American mathematics education there is considerable debate about the status of certain kinds of knowledge. One side is accused of interesting itself solely in "skill-drill" and computation, the other of interesting itself solely in "fuzzy math", where anything goes as long as it is in the right general vicinity. Consider instead the following description: all school learning is an alternation of *savoirs* and *connaissances*. Isolated parts are acquired as *savoirs* connected by *connaissances*. Without the *connaissances*, the *savoirs* have no context and are swiftly mixed or lost. Without the *savoirs*, the *connaissances* are more touristy than useful. Imbedded in *connaissances*, *savoirs* can develop gradually into a solidly connected chunk—in fact, a *savoir*, which is then available to be set into a wider *connaissance*. Thinking this way then provides a tool for contemplating another of the current hazards of mathematics education: assessment, which is a clear need, but a thorny issue. And one of the causes of its thorniness is that all that traditional standardized tests can measure well is *savoirs*. The state of a student's *connaissances* is visible to the teacher if enough time in the classroom can be devoted to the kind of activity where *connaissances* are built and used. But an overemphasis on visible, "testable" knowledge leads to attempting to teach the *savoirs* without the *connaissances* to hold them together and carries with it the danger of damaging the whole fabric of the learning, leaving the learner with a collection of isolated facts and no understanding of how they are connected.

It should by now be clear that a casual treatment of the *savoir/connaissance* distinction would be a serious error. On the other hand, finding a solution is not a trivial pursuit. Some researchers distinguish between "knowledge" and "understanding", but these do not have the universally accepted distinction of the French words. One solution would be simply to transfer the words, untranslated, as we have done with the similarly untranslatable "*Didactique*". On the other hand, that would be cumbersome and, given the conjugations of the verbs, both obscure and distracting. Past efforts have included use of "know-how" and "a knowing", but neither has proved very satisfactory. We (Brousseau and Warfield) therefore converted to using the following convention: Some of the time we use "to know", "knowledge" and "a piece of knowledge" (the latter for the singular form of either noun in French), but when there is a need to make the distinction we attach the prefix "c-" when the word comes from a form of "*connaître*" and "s-" when it comes from a form of "*savoir*". We hope in this way to achieve the best available balance of accuracy and readability. [2013 Note: As it turned out, using s-knowledge and c-knowledge was even more cumbersome than using the words untranslated, so in the course of the intervening years we have reverted to that solution, with the occasional substitution of "reference knowledge" for savoir.]

14. Metacognitive slippage

This consists of substituting for a concept a model of it, or a nearly, but not quite completely, equivalent concept. An example of the former would be substituting for the formal definition of continuity the model of "something whose graph can be drawn without lifting the pencil from the paper." In the latter category come the frequently observed use of equality and equivalence as synonyms.

15. Didactical transposition

This is another form of slippage, but a necessary one: mathematical concepts in their full glory are rarely in a form that it is even possible, much less desirable, to teach. An absolute necessity within the dictates of mathematical education is to carry out a transposition of concepts into a teachable form. Equally a necessity is that we keep the issue under close surveillance so as to forestall the clear danger of distorting the mathematics in the process.

16. The Topaze effect

This is a cultural reference that is extremely useful within the field for the amount that it reflects in a quite compact format, and is extremely tough on the translator for both cultural and linguistic reasons. Over the past decade I have made a sequence of translations, each time feeling cleverer than the last, and each time discovering afterwards some nuance that has nonetheless disappeared. This time I shall rely more on description and less on translation.

The reference stems from a well-known comedy by Marcel Pagnol whose main character starts the play as a thoroughly unsuccessful classroom teacher named Topaze. He is dictating a sentence to a somewhat unintelligent and thoroughly reluctant young boy. The objective is for the boy to write it correctly, which entails not merely spelling, but grammar recognition. The sentence is "Des moutons étaient réunis dans un parc." ["Some sheep were gathered in a field."] The major challenge stems from the fact that the *s* which makes "moutons" plural is silent, and further-more "étaient" (plural) is pronounced exactly the same as "était" (singular). In the course of a sequence of repetitions, Topaze gives thicker and thicker hints, up to actually pronouncing the *s* and the ent ("des moutonsses étai-hunt...") thereby totally astonishing the lad, but not, in fact, inducing him to write the desired forms.

The issue, when the Topaze effect is cited, is the process of changing the initial aim, which was for the student to demonstrate an understanding of the grammatical issue of pluralizing nouns and verbs, into the purely mechanical one of getting an *s* onto the piece of paper. By the end of the scene Topaze is clearly on the brink of grabbing the pencil and writing it himself, thereby completing the process of remov-ing any learning from the situation.

17. The Jourdain effect

This is another cultural reference, but a considerably more internationally recogniz-able one than the preceding. Jourdain is the protagonist in Molière's *Bourgeois Gentilhomme*, and a model of pretentious ignorance. His tutor cashes in on both of these characteristics and keeps him very happy by discovering brilliance in every humdrum detail. The classic example is the lesson from which Jourdain emerges totally full of himself because he has discovered that he is speaking in prose.

Bibliography

Brousseau, G.: Theory of Didactical Situations in Mathematics. Kluwer Academic, Dordrecht (1997)

Brousseau, G.: Les Mathématiques du Cours Préparatoire, Fascicule 1. Dunod (1965)

Brousseau, G.: Processus de Mathématisation, La mathématique à l'école élémentaire, pp. 428–442. Association des professeurs de mathématiques. Paris (1972)

Brousseau, G., Warfield, V.: The case of Gaël. J. Math. Behav. **18**(1), 7–52 (1999)

Brousseau, G., Brousseau, N., Warfield, V.: An experiment in the teaching of statistics and probability. J. Math. Behav. **20**(3), 363–411 (2001)

Brousseau, G., Brousseau, N., Warfield, V.: Rationals and decimals as required in the school curriculum Part 1: Rationals as measurement. J. Math. Behav. **23**, 1–20 (2004)

Brousseau, G., Brousseau, N., Warfield, V.: Rationals and decimals as required in the school curriculum Part 2: From rationals to decimals. J. Math. Behav. **26**, 281–300 (2007)

Brousseau, G., Brousseau, N., Warfield, V.: Rationals and decimals as required in the school curriculum. Part 3: Rationals and decimals as linear functions. J. Math. Behav. **27**, 153–176 (2008)

Brousseau, G., Brousseau, N., Warfield, V.: Rationals and decimals as required in the school curriculum. Part 4: Problem-solving, composed mapping and division. J. Math. Behav. **28**, 79–118 (2009)

Perrin-Glorian, Marie-Jeanne: Théorie des Situations Didactiques: Naissance, Développement, Perspectives. In: Vingt Ans de Didactique des Mathématiques en France, Pensée Sauvage (1994)

Pirie, S.E.B., Kieren, T.E.: Growth in mathematical understanding: how can we characterise it and how can we represent it? In: Cobb, P. (ed.) Educational Studies in Mathematics, **26**(2–3), 165–190 (1994)

V.M. Warfield, *Invitation to Didactique*, SpringerBriefs in Education 30,
DOI 10.1007/978-1-4614-8199-7, © The Author(s) 2014

Index

V.M. Warfield, *Invitation to Didactique,* SpringerBriefs in Education 30,
DOI 10.1007/978-1-4614-8199-7, © The Author(s) 2014

S

Situation, 66
    of action, 9, 10, 19, 61, 66
    of communication, 66
    of formulation, 10, 66
    fundamental, xv, 21, 37–38,
        52, 67
    of institutionalization, 12–13, 66
    paint pots, xiii, xiv
    vs. situation, 11

    the thickness of a sheet of paper, 22–27
    of validation, 10, 66
S-knowledge, 68–69
Statistics, 37

T

Theory of situations, xv, 6, 9, 12, 15,
    21, 37, 56, 65, 67, 68
Topaze effect, 70